Life and Process

Process Thought

Edited by
Nicholas Rescher, Johanna Seibt, Michel Weber

Volume 26

Life and Process

Towards a New Biophilosophy

Edited by
Spyridon A. Koutroufinis

DE GRUYTER

ISBN 978-3-11-034326-7
e-ISBN 978-3-11-035259-7
ISSN 2198-2287

Library of Congress Cataloging-in-Publication Data
A CIP catalog record for this book has been applied for at the Library of Congress.

Bibliographic information published by the Deutsche Nationalbibliothek
The Deutsche Nationalbibliothek lists this publication in the Deutsche Nationalbibliografie; detailed bibliographic data are available in the Internet at http://dnb.dnb.de.

Printing: CPI buch bücher.de GmbH, Birkach
♾ Printed on acid-free paper
Printed in Germany

www.degruyter.com

To the memory of Reiner Wiehl (1929-2010), the spiritus rector of German Whitehead research.

Table of Contents

Preface

Isaac Newton famously remarked that if any of us can claim to see further "it is by standing on the shoulders of giants." We present here a new biophilosophy, our vista achieved by standing on the shoulders of Alfred North Whitehead, one of the greatest philosophers and scientists in history.

In deciding that we should present in a single volume our ideas about the relevance of Whitehead's metaphysics for 21st century biosciences, I was inspired by conversations with American and European colleagues whom I met at conferences and workshops on process philosophy over the past decade. Editing this volume has advanced my understanding of how Whitehead's metaphysics can become the philosophical foundation for a biology that surpasses the machine-metaphor prevalent in biology today.

One of the main challenges of this century is finding ways to describe biological phenomena at all scales as persisting processes rather than as systems of fixed parts with specific functions, a very demanding challenge given the dominance of reductionistic and mechanistic ontologies in academic biology.

The book addresses subjects that, at first, may seem to be widely divergent. I hope that after having read this volume the reader will agree that Whiteheadian metaphysics offers the ideal background for considering these topics as mutually interdependent dimensions of a new and truly integrated biophilosophy.

We dedicate this book to Reiner Wiehl, a pioneer and strong campaigner for our new Whiteheadian perspective. Wiehl, who was a research fellow and the assistant to the eminent hermeneuticist Hans-Georg Gadamer, was a tenured professor for philosophy at the University of Heidelberg until his retirement in 1997. He was one of the advisors of my habilitation thesis *Organismus als Prozess* (*Organism as Process*) which I completed in the Institute for Philosophy of the Technical University of Berlin in 2009. I owe tribute to Reiner Wiehl for many deep insights into the significance of Whitehead's thought. He was working on a chapter for this volume when he passed away.

We miss him and wish he were here with us today to celebrate this book's publication.

Spyridon A. Koutroufinis

University of California, Berkeley,
October 2013

Acknowledgments

I have had to edit this book in a language that is not my native tongue. I would not have overcome the many obstacles without the great support of my colleagues Terrence Deacon, Robert Valenza, Andrew Packard, and Jonathan Delafield-Butt.

List of Contributors

John B. Cobb, Jr., PhD, Professor Emeritus,
Co-director of the Center for Process Studies, Claremont School of Theology, Claremont CA.

Jonathan T. Delafield-Butt, PhD,
Faculty of Humanities and Social Sciences, University of Strathclyde, Scotland, UK.

Joseph E. Earley, Sr., PhD, Professor Emeritus,
Department of Chemistry, Georgetown University, Washington, D.C.

Gernot Falkner, Dr., Professor,
Cell Biology Department, University of Salzburg, Austria. Formerly Researcher at the Institutes of Molecular Biology (Salzburg) and Limnology (Mondsee) of the Austrian Academy of Sciences.

Renate Falkner, Dr.,
Formerly Researcher at the Institutes of Molecular Biology (Salzburg) and Limnology (Mondsee) of the Austrian Academy of Sciences.

David R. Griffin, PhD, Professor Emeritus,
Co-director of the Center for Process Studies, Claremont School of Theology, Claremont CA.

Pete A.Y. Gunter, PhD, Professor Emeritus,
Department of Philosophy and Religion Studies, University of North Texas, Denton, TX.

Spyridon A. Koutroufinis, Dr., Privatdozent,
Institute of Philosophy, Technical University of Berlin, Germany.

Barbara Muraca, Dr., Post-Doc.,
Institute of Sociology, Friedrich-Schiller University, Jena, Germany.

Andrew Packard, DSc.,
Formerly Naples Zoological Station, Reader in Physiology, Edinburgh University, Scotland, UK, and Professor of Zoology, Università di Napoli Frederico II, Naples, Italy.

Henry P. Stapp, PhD, Theoretical Physicist,
Lawrence Berkeley National Laboratory, University of California, Berkeley CA.

Robert J. Valenza, PhD, Professor,
Department of Mathematics, McKenna College, Claremont CA.

Introduction: The Need for a New Biophilosophy[1]

SPYRIDON A. KOUTROUFINIS

Alfred North Whitehead is often regarded as the most original innovator of 20th century philosophy of nature and metaphysics. In recent decades a number of leading theoretical physicists have introduced ground-breaking new perspectives on fundamental issues of physics on the basis of his process philosophy. In contrast most biologists have not seriously questioned the Cartesian metaphysics of 19th century classical physics and only just begun thinking about possibilities of overcoming it. This book aims to contribute to the foundation of a new direction in biophilosophy which goes beyond many of the core metaphysical assumptions of contemporary mainstream biology. All of the co-authors of this volume treat central metaphysical questions about the nature of life from the perspective of Whitehead's process philosophy. These questions are crucial for the biosciences, but cannot be addressed by them since they touch on metaphysical issues.

In order to show the plausibility and the sense of this enterprise, first I will explain why I believe it is necessary to differentiate between biophilosophy and the philosophy of biology. Second I will review some of the shortcomings of today's biology and philosophy of biology and demonstrate how a biophilosophy grounded in a process-oriented metaphysics can overcome them. Third, I will provide a summary of Whitehead's process ontology, emphasizing those fundamental ideas from this paradigm that play essential roles in the present book. Finally, I will briefly describe the main ideas presented in subsequent chapters.

[1] I gratefully acknowledge the editorial help and critical remarks of Terrence Deacon, Robert Valenza, Andrew Packard, and Jonathan Delafield-Butt.

1. Biophilosophy and Philosophy of Biology

Philosophy of biology is a discipline which was founded in the early 1970s by the efforts of Michael Ruse (1973) and David Hull (1974), but which had also had some precursors (Beckner 1959). The best-known representatives of this discipline, which has become especially established in the Anglo-American world, are theoretical biologists and philosophers.[2] Many authors also refer to the philosophy of biology as "biophilosophy". However, I do not think that these two labels should be used synonymously. I describe "biophilosophy" as a philosophic tradition existing since antiquity which includes a set of very different, heterogeneous philosophic considerations of life. From this point of view, philosophers of biology constitute only one subgroup within the broader category of biophilosophy, even though they are arguably the most influential group today.

There are two reasons why I suggest making this distinction between biophilosophy and philosophy of biology and consider the latter to be included in the former: First, considering biophilosophy to be the metaphysically more broadly conceived field allows one to point to the relevance of the works of philosophers like Aristotle and Kant to current biosciences without characterizing them as "philosophers of biology", which could be somewhat misleading given that the term "biology" was only introduced at the beginning of the 19th century when this discipline was founded. Second, in contrast to most scholars who understand themselves as philosophers of biology, and who, in their reflections about matter and causality, almost never contravene the basic metaphysical framework dictated by today's mainstream biology, the philosophical presuppositions of the biophilosophers follow very different metaphysical systems. This being said, however, it is important to note that the borders between both fields are fluid

The most important Western thinkers of biophilosophy who will remain relevant in its future are Aristotle and Kant. Other philosophers and scientists with considerable influence on biophilosophy are William Harvey,

[2] Some of the most influential contributions to philosophy of biology have been provided by Francisco Ayala, Theodosius Dobzhansky, John Dupré, Steven Gould, Paul Griffiths, Richard Lewontin, Huberto Maturana, Ernst Mayr, Susan Oyama, Alexander Rosenberg, Elliott Sober, Kim Sterelny, and Francisco Varela.

Gottfried Wilhelm Leibniz, Wolfgang von Goethe, Carl Gustav Carus, Gustav Theodor Fechner, Charles Darwin, Ernst Haeckel, Friedrich Nietzsche, Henri Bergson, Hans Driesch, Alfred North Whitehead, Charles Sanders Peirce, Jakob von Uexküll, Kurt Goldstein, Georges Canguilhem, Viktor von Weizsäcker, Adolf Portmann, Hans Jonas, Michel Foucault, and Gilles Deleuze. Recently, many contemporary bioscientists have provided new conceptions of organism, evolution, and consciousness which clearly transcend the frame of mainstream philosophy of biology.[3]

All forms of biophilosophy, including philosophy of biology, deal with questions that arise out of biology but which biology cannot answer. The central question revolves around our understanding of the concept of "life" – its meaning or semantic extension. In 20th century biophilosophy, this concept has a wide spectrum of connotations. On one level, "life" refers to the totality of processes which occur in any given physical entity that is described as an "organism". On another level, this concept refers to sets of such entities. So "life" often refers to a group of organisms of the same species (e.g., an animal colony) or to the interacting species of an ecosystem or even to the entire biosphere. Frequently "life" means all organisms which have come into being since the appearance of the first cell on the early earth, with some bioethicists even using this concept to refer to all future organisms. Sometimes the concept of "life" also includes hypothetical biological developments which could occur outside of the earth (exobiology), thus going beyond the spatiotemporal limits of evolution on earth. These different facets of the term "life" are present in virtually all of the forms of contemporary biophilosophy. The only really controversial question is whether real or potentially real products of the "Artificial Life" (AL) project, i.e., computer simulations of organisms and ecosytems (e.g. Tierra or Daisyworld), "intelligent" robots, or future self-reproducing automata (which would be physical entities rather than computer simulations), should be included in the category of "life". Proponents of the so-called "strong AL" follow John von Neumann's position that life is a specific form of dynamics which can be abstracted away from any particular medium (1966). Interestingly enough, some postmodern biophilosophers, although their methods have nothing in

[3] Kauffman 2008, 2002, 2000; Deacon 2012, 2006; Hameroff 2007, 2003; Hameroff and Tuszynski 2004

common with the analysis methods of the natural sciences, support the strong AL project insofar as they often include real and possible future products of the AL project in the phenomenon of "life".

The differences between the varying forms of biophilosophy become clearer in the context of the question about the *nature* or *essence* of life. Here, too, biophilosophers influenced by Deleuze and other postmodern thinkers hold a distinctive position. They reject the idea that life has an "essence", underscoring instead the incomprehensibility of the phenomenon, namely its tendency to transcend any characteristics (Thacker 2005). Other biophilosophers, who do not follow postmodernism, consider the question of the *nature* or *essence* of life to be pivotal. Their answers reveal the basic metaphysical ideas with which they operate, which may vary considerably between different thinkers.

Today's philosophy of biology is built upon metaphysical assumptions about matter, causality, and mental agency (and their respective places in the cosmos) that are substantially different from the metaphysical assumptions of Aristotelian, Jonasian, Whiteheadian and other biophilosophy. Most philosophers of biology follow the metaphysical principles of classical physics, of course in a version that is expanded to include the idea of dynamical systems, which include the theories of complexity, self-organization, and chaos. For the purposes of this volume, the following basic metaphysical principles are important, since they are explicitly rejected by biophilosophers who have a process-metaphysical or other perspective:

- Mental activity is inseparably connected to brain activity. Plants, simple multi-cell organisms and single-cell organisms do not experience anything. The ability to experience arose relatively late in the history of evolution and is reducible to complex physicochemical patterns of activity in neural systems.
- Mental or other factors which cannot be reduced to physicochemical processes possess no causal relevance for biological occurrences. Mental states are irrelevant to ontogenesis[4] and evolution, even though they may

[4] Aristotle, on the other hand, argues that mental factors have an effect on and form matter, and makes them the foundation of his teleology (Koutroufinis, this book, section 2.3).

appear to be an important evolutionary factor due to their role in partner selection.[5] In reality, however, they are causally irrelevant epiphenomena which can be reduced to the interaction of neural, genetic, and signal networks.

- All processes in an organism can be understood as arising from the interactions of material entities that are strictly localized spatio-temporally. Ideas in quantum physics such as the non-local entanglement between elementary particles are generally not considered to be relevant in biology (including brain physiology). For this reason the ideas of classical physics about matter and determinism – again, in a version expanded by the theories of complexity, self-organization, and chaos – are sufficient for understanding the causal relations in biological processes (e.g., signal processes as well as metabolic, genetic and neural networks).

The focus on the classical-physical notion of matter and causality means that the last basic assumption directly supports the first. This is the case because – in stark contrast to quantum theory – classical physics (and with it also theories of complexity, self-organization, and chaos) excludes any form of subjectivity from physical causality.

It should be noted that these assumptions do not say anything about the *methodology* of biology, as they are ontological and metaphysical assumptions about the nature of the matter and causality of organisms. Philosophers of biology are often skeptical when methods from physics are carried over and applied to biology.[6] As metaphysical assumptions, however, they indicate the *Weltanschauung* of modern biology. The vast majority of biologists are convinced that life can be explained "naturalistically". In this

[5] Darwin's concept of "sexual selection", which is central in his work *The Descent of Man, and Selection in Relation to Sex*, was founded on the idea that experience and appreciation of beauty are fundamental in the mate selection of even simple animals like insects (1989, 304). Contemporary treatises on sexual selection avoid using any terms which could be associated with mental activity and especially with experience (Zahavi 1975, 1997).

[6] For example, one of the problems often discussed in philosophy of biology is whether there are causal factors at work in evolution that can be thought of as being similar to physical forces.

case "naturalism" is hardly ever defined, even if the author explicitly describes him or herself as a "naturalist". The naturalism of most biologists is usually a particular form of physicalism which does not even consider notions of matter and causality that have long been established within quantum physics. This means that, in most cases, the metaphysics of the physicalistic naturalism of contemporary biology is the metaphysics of physics *before* the development of quantum theory. The most central characteristic of this naturalism is that there must be no relying on the so-called "supernatural" in scientific explanations. This usually includes not only ideas of God but also everything which cannot be understood using the methods of physics (and chemistry); in other words, this also includes interiority or phenomenal qualities (qualia) of experience and other mental phenomena. From this point of view, since nature is considered to be purely the totality of material and energetic phenomena, mental agency can only be considered as "supernatural".

Whereas most philosophers of biology operate within this metaphysical framework dictated by contemporary mainstream biology, the basic assumptions of some biophilosophers break these barriers. However, this does not mean that they reject naturalism. These biophilosophers can be seen as representatives of an expanded form of naturalism – a *liberal naturalism*. Over the past few years several philosophers have introduced this new form of naturalism.[7] It allows mental states, such as phenomenal qualities, as aspects of natural entities and ascribes ontological relevance to abstract, modal, moral, and intentional entities.[8] John Dupré, a philosopher of biology and decisive critic of scientific monism (2004), is one of the supporters of liberal naturalism (2010). Dupré's work shows that there can be a continuous transition between philosophy of biology and the broader field of biophilosophy. His main epistemological and ontological positions could be advocated by biophilosophers who follow Whitehead's philosophy of life. In the past the borders of philosophy of biology within biophilosophy have often shifted. This will probably happen again in the future. Those borders depend on the main ontological assumptions of phi-

[7] De Caro and Macarthur 2010a; De Caro and Voltolini 2010, 75-82; McDowell 2004
[8] De Caro and Macarthur 2010b, 12

losophers of biology which are influenced by the prevalent metaphysical paradigm of main stream biology that might change.

Just why most contemporary philosophers of biology seem unwilling to follow such a liberal naturalism can be explained by recognizing the inordinate historical influence of one branch of biology, namely theoretical biology. This discipline began in the early 20th century with the works of Johannes Reinke (1901), Jakob von Uexküll (1909, 1920), and Julius Schaxel (1919) aimed to develop a philosophically consistent foundation for biology. But in the 1920s, Alfred Lotka (1925) and Vito Volterra (1926, 1931) developed mathematical models of population dynamics and became the forerunners of the systematic mathematization of theoretical biology, which began in earnest in the 1930s with the works of Ludwig von Bertalanffy (1932). Important contributions to the founding of mathematical theoretical biology were also made by Nicolas Rashevsky (1938, 1940), Erwin Schrödinger (1944) and Alan Turing (1952). With the development of theories of nonlinear dynamic systems and the derivative concepts of self-organization, chaos and complexity following the pioneering contributions of William R. Ashby (1962)[9], Heinz von Foerster (1960, 1962), Ilya Prigogine (1967, 1968)[10], Hermann Haken (1973, 1983) theoretical biology became a mathematical discipline[11] – which is why it is often referred to as "biomathematics". As a result, the originally wide range of topics became much more limited. In today's institutes of theoretical biology, it is mainly mathematical models and computer simulations for processes from evolution theory, developmental biology, ecology, neurobiology and epidemiology that are tested. In other words, it seems that only the branch of theoretical biology which can be traced back to Bertalanffy has survived. But even here, several of Bertalanffy's important philosophical intuitions have disappeared from view.[12] In this way the change of theoretical biology to

---

[9] The term "self-organizing system" was introduced into scientific discourse by Ashby 1947.

[10] Prigogine and Nicolis 1967, Prigogine and Lefever 1968. See also: Glansdorff and Prigogine 1971, Nicolis and Prigogine 1977, Prigogine and Stengers 1984

[11] Kauffman 1995, 1993; Murray 1993; Goodwin 1994; Goldbeter 1997; Noble 2006

[12] In his book *Problems of Life* (first published in German 1949), Bertalanffy speaks of a new "non-quantitative" mathematics (also "Gestalt mathematics") for biology, in

biomathematics has led to a large gap in regards to theoretic reflections about the foundations of biology as a scientific discipline. Today, this gap is increasingly being closed by philosophy of biology, which addresses such topics as the relation of biology to physics and chemistry; the concepts of natural selection and adaptation; the level or levels at which natural selection acts (genes, organisms or groups of organisms); the concept of the gene; the meaning of teleology, purpose, and function; the relationship between micro- and macroevolution; the nature of biological species; the emergence of humankind in evolution; the role of genetic factors in human behavior; the relation of biology to ethics; ecology and the concept of ecological diversity; and the conflict between evolution and theism.[13] The wide range of these topics shows that the borders between theoretical biology and philosophy of biology are fluid; this also explains the very close connection between the latter and the metaphysics of mainstream biology.

Due to its long development since classical antiquity, biophilosophy is able to exhibit different naturalistic understandings of life in general and of the organism in particular which transcend the physicalistic metaphysics of most biologists. For each of these directions several aspects are essential. Aristotle was convinced of the fact that factors of a non-material nature regulate the growth and self-preservation of an organism.[14] Contemporary biophilosophers have further developed this idea. Hans Jonas, who is strongly influenced by Aristotle's view of teleology, stresses the interiority and freedom of every organism (2001). Adolf Portmann describes the "Tiergestalt" (animal-gestalt) as an expression of interiority, which he understands as the richness of experience of an animal organism (1960). Prior to this, in his work *The Descent of Man*, Charles Darwin had established the notion of sexual selection, and stated that, in this form of selection, *acts of experience* occupy a central role, even for non-complex animals.

---

which not quantity but rather the idea of form or order would come to occupy the central role (1952, 159-160).

[13] Rosenberg and McShea 2008, Hull and Ruse 2007, Griffiths 1992, Rosenberg 1985

[14] *On the Soul* II and *Physics* II

2. The need for a process-metaphysical biophilosophy

The self-limitation of contemporary biosciences and philosophy of biology to the physicalistic metaphysical principles about causality, matter, and mental agency mentioned above causes serious difficulties in understanding essential aspects of life.

Firstly, the reduction of organismic processes (e.g. signal, metabolic, and genetic processes) to the interaction of spatio-temporally strictly localized material entities (atoms and molecules) the nature or essence of which is fixed (substantialism) makes it impossible to understand the self-organization of real organisms. Theorists of self-organization, dynamical systems, and complexity as well as systems-biologists (who apply these theories to biology) face fundamental limitations in explaining the dynamics of a whole organism, e.g. a cell, as mutual interrelatedness of a big number of organismic processes without making unrealistic pre-assumptions.[15] The theory of self-organization can sufficiently describe non-living dynamical systems of physics. But in order to provide persuasive models of organismic self-organization it needs first of all to develop a well defined conception of *self*.[16] Organismic self has to be conceived of as a form of dynamics which transcends physicalistic metaphysics insofar as it autonomously makes an important distinction which constitutes the organism – it determines its own boundaries which necessarily defines its own physical surroundings or non-self. Physicalistic metaphysics does not provide a sufficient basis for explaining how a complex entity manages to differentiate itself from its physical surroundings. Physical systems require that an external factor defines their limits, thus making the distinction between what belongs to the system and what to its surroundings. In contrast all organisms, even the simplest bacterium, make this essential distinction on their own. Organismic self-maintenance and self-creation requires that the living self is able to sustain the flux of energy and matter from its physical surroundings by its own dynamics. This means that a living being must be able to interpret what part of its physical surroundings is relevant to it. It separates the part of its non-self

---

[15] Koutroufinis in this book (section 1.3, 1.4); 2013, 323-327; Deacon and Koutroufinis (forthcoming); Koutroufinis and Wessel 2013

[16] Deacon and Koutroufinis (forthcoming), Koutroufinis 1996

that is relevant to its own life from the irrelevant part. Those features of the physical surroundings that are represented with respect to the maintenance of internal organismic dynamics constitute what Jakob von Uexküll termed *Umwelt*. Self and *Umwelt* are two sides of the same coin. This indissoluble connection is not comprehensible from the point of view of a physicalistic metaphysics, since non-living physical systems have neither *Umwelt* (they have only externally set surroundings) nor self.

Secondly, the autonomous constitution of self and *Umwelt* are indissolubly connected to the end-directedness or teleology of organismic dynamics. Given the ignorance of classical physics for *Umwelt* and self it is not surprising that one of its most important principles since its foundation in the 17th century is that teleology has to be totally excluded from the study of nature. However, in the last century a new kind of teleological thinking was reintroduced by cybernetics,[17] evolution theory,[18] and theory of self-organization[19].[20] Teleology is purposeful end-directedness. As such it requires both memory and anticipation of future events to some degree. Without these faculties organismic dynamics would not be able to adjust non-simultaneous processes in such a way that early processes provide supportive conditions for the occurrence of later ones as happens in all healthy organisms.

Thirdly, physicalistic metaphysics and Neo-Darwinism cannot explain the growth of interiority and the progressive complexification of experience in evolution. Animal experiences are usually teleologically organized since they are directed to a certain end-state. For example the olfactory experience of an animal which finds its nourishment by following a gradient of odor intensity to its source is not an end in itself but serves for the sustainment of the living self. In other words, organismic teleology is not only a purposeful end-directedness but also an intentional one. But within the framework of physicalistic reductionism phenomenal qualities (qualia) – such as the experience of joy, hunger, beauty, fear, sympathy, and antipa-

---

[17] Rosenbleuth, Wiener , and Bigellow 1943

[18] Deacon 2012, 421-462; Ariew 2007; Bedau 1998; Mayr 1991, 56; Brandon 1990, 188; Ruse 1988, 44; Hull 1974, 103

[19] Christensen 1996

[20] For more details about the reintroduction of teleological thought in natural sciences see Koutroufinis 2013.

thy – are understood as epiphenomena of complex blind (not mental) physicochemical processes. Epiphenomenalism and materialistic eliminativism force biologists to reject the causal relevance of the interiority of experience to the behavior even of higher animals. When viewed from this perspective, using mental factors to explain the striving of organisms for nourishment, safety, procreation, or getting to a particular place makes no sense. However, from an evolutionary point of view the rejection of the causal relevance of qualitative phenomena raises an unanswerable question: if qualia have no function in physical reality why has the ability of qualitative experience been positively selected by evolution instead of having disappeared long ago? Obviously it is impossible to think seriously about the interiority of experience without having a sufficient theory of self, i.e., one which includes the ability of phenomenological experience.

Fourthly, following the main metaphysical assumptions of classical physics reduces matter to spatio-temporally strictly localized particles and causality to determinism (which includes deterministic chaos). This precludes essential quantum-theoretical ideas, like non-local entanglement and indeterminism, from theory of life. Recently, however, a number of publications in leading scientific journals demonstrate that in their long evolution organisms have found ways to utilize quantum phenomena.[21] Today there is a lot of evidence that quantum coherence plays an important role in metabolic and neuronal processes.[22] Quantum biology is a rapidly emerging discipline. Its foundation, however, can be traced back to pioneers of quantum theory, like Niels Bohr, Werner Heisenberg, Walter Heitler, Walter Elsasser, and Pascual Jordan.[23] Quantum biology is of great philosophical relevance. As it is based on quantum theory, which has introduced a new view of the subject-object relation, quantum biology supports the establishment of a new understanding of the role of subjectivity and freedom in the theory of organism, neurobiology, and evolution theory.

Whiteheadian metaphysics allows one to avoid all four shortcomings of physicalistic biology and philosophy of biology. We shall see in the fol-

[21] Collini et al. 2010, Engel et al. 2007, Lee et al. 2007

[22] Davia, 2006, 268ff; King 2006, 439; 1996, 208ff; Mershin et al. 2006, 103ff; Hameroff and Tuszynski 2004, 28ff.

[23] Bohr 1933[1990]; Heisenberg 1984[1990]; Heitler 1976[1990]; Elsasser 1990, 1982; Jordan 1947, 1972[1990]

lowing section that it is ideally suited to creating a biophilosophy that integrates the essential aspects of self, *Umwelt*, experience, and macroscopic quantum coherence of organismic processes as these are also essential aspects of Whiteheadian ontology.

3. The foundation of Whitehead's metaphysics

The mathematician, physicist and philosopher Alfred North Whitehead (1861-1947) succeeded in becoming the best known representative of *process philosophy*, a discipline which arose around 1900 and is now accepted as an independent philosophical tradition. Genuine process philosophical ideas can be found in the writings of Friedrich Nietzsche, Charles Sanders Peirce, Henri Bergson, William James, Samuel Alexander, John Dewey, Nicholas Rescher, and of course in the works of the Whitehead-scholar Charles Hartshorne. In his process philosophy, Whitehead introduces a new kind of teleological thinking which is not based on any type of substance ontology. Process philosophy provides interesting ways to avoid the problems of vitalism and allows the principal problems inherent in an understanding of life based on contemporary natural sciences to be overcome.

The purpose of the following summary of the basics of Whitehead's ontology is to introduce readers to the common themes which run through the contributions in this book. For Whitehead's philosophy in general we refer you to the extensive secondary literature.[24]

3.1 Actual entities: the final facts are processes

The basic premise of all process philosophies is that all primary entities in the universe are processes. Everything which persists in space time is understood as the result of sequential manifestations of interconnected and interrelated processes. Whitehead calls the most elementary, indivisible facts of reality *actual entities* or *actual occasions*. In Whitehead's meta-

---

[24] Leclerc 1975; Sherburne 1961; Lowe 1966, 1985, 1990; Christian 1967; Emmet 1981; Lango 1972; Kraus 1979; Fetz 1981; Hampe 1998; Rust 1987; Sayer 1999

physics, this ontological category takes over the role of the "primary substance" in Aristotle's philosophy. However, in virtue of being processes, actual occasions cannot be appropriately understood if any of the possible meanings of "substance" in philosophical, scientific, and every day language are applied to them.

> "'Actual entities' – also termed 'actual occasions' – are the final real things of which the world is made up. There is no going behind actual entities to find anything more real. They differ among themselves: God is an actual entity, and so is the most trivial puff of existence in far-off empty space. [...] The final facts are, all alike, actual entities [...] The notion of 'substance' is transformed into that of 'actual entity'" (1979, 18f.).

In his main work, *Process and Reality*, Whitehead often characterizes actual entities as "*res vera* in the Cartesian sense of that term" (1979, XIII) which means things that they are "real and true" (ibid., 74). But other than Descartes, who "retained in his metaphysical doctrine the Aristotelian dominance of the category of 'quality' over that of 'relatedness'", Whitehead anchors his central argument on the basic assumption that "'relatedness' is dominant over 'quality'" (ibid.).[25]

3.1.a. Physical-mental bipolarity

A second basic premise of Whitehead's metaphysics is the assumption that actual entities or actual occasion are inseparable *physical-mental unities* – one of the key ways in which he shows an affinity to Leibniz. It is in this way that Whitehead reacts to what he calls the "bifurcation of nature", the separation of nature into mind and matter that has played dominant role in philosophy and science since the 17th century. According to Whitehead, this is the only way to integrate even the most simple act of experience into a nature devoid of "mind". That increasing the neuronal complexity of a dynamics based solely on efficient causation should lead to the sudden addition of an inner dimension of experience is something that Whitehead considers – justifiably – incomprehensible. Therefore, he conceives of the

[25] See section 3.1.b of this introduction.

actual entities as *processes of experience*, which he calls subjects of their own experienced immediacy:

> "The actualities of the Universe are processes of experience, each process an individual fact" (1967, 197).
> "An actual entity is called the 'subject' of its own immediacy" (1979, 25).

Actual entities are physical-mental bipolar unities. They are entities endowed with subjectivity that are always related to and can also generate things existing physically in space-time (see 3.2.b). While Whitehead's process philosophy is based on a *pansubjective* ontology it cannot necessarily be categorized as panpsychism.[26]

For his part, Whitehead does not tire of arguing against equating mental activity with consciousness. Like Aristotle and Leibniz, Whitehead explains that the term "mental" is much more comprehensive than "conscious", as only very few mental phenomena can be classified as possessing consciousness. Nearly all actual entities are merely *protomental* events and as such they are not "conscious".[27] Different processes are configurations of widely variable type and may exhibit any number of grades of consciousness, including a complete lack thereof, depending on their complexity. *But all processes are complexes of experience.* Thus, the idea of experience plays a much greater role in Whitehead's concept of a subject than does the idea of consciousness.

Whitehead's metaphysics can also be assigned to liberal naturalism. Only those philosophers who subscribe to the narrow concept of physicalistic naturalism, described above, are surprised to hear that Whitehead's philosophy of nature is a special kind of naturalism.[28]

---

[26] According to Wiehl Whiteheadian pansubjectivism is a revised panpsychism (1990, 217).
[27] Aristotle makes this clear in *Physics* II, 199 b26-30, Leibniz in *Monadology*, §19 and in *Principles of Nature and Grace*, §4, and Whitehead in *Process and Reality* (25, 53, 56, 139, 280), *The Function of Reason* (16, 32, 33), and *Adventures of Ideas* (180).
[28] Griffin shows in what respect Whitehead's philosophy of nature is a specific kind of naturalism (this book).

3.1.b. Internal relationality

One of the main reasons for Whitehead to depart from the old metaphysics of substance was that, in his opinion, "[t]he relations between individual substances constitute metaphysical nuisances: there is no place for them" (1979, 137). The Cartesian substance as something that "exists in such a way that it doesn't depend on anything else for its existence"[29] is conceived of as being self-sufficient. As such it requires no relation to anything else in order to exist. Whitehead explicitly distances himself from this conception of substance (ibid. 59). He also thinks that the concept of the actual entity is inconsistent with Aristotle's "primary substance" in his early work, which is "neither predicable of a subject nor present in a subject".[30] However, Whitehead clearly sees the conceptual difference of Cartesian substance from Aristotelian "primary substance" (ibid. 137-138).

The actual entities are subjects, but not in the sense of the classical metaphysical idea of subjectivity as a feature of a substance. As a processsual subject is not a substance, it cannot relate to its own experiences as a timeless carrier, the essence of which is not changed by its experiences. Therefore, Whitehead's way is only possible if one does not separate the *essence* of the processual subject from its experiences. He conceives of the actual entity (that is, the processual subject) as a totality of experiences that grows together to form a whole. The source of these experiences cannot be found only within the subject, as this is after all not the window-less monadic substance as described by Leibniz in his *Monadology*. The processsual subject must be able to experience things through the "window" of its relation to reality that consists of the totality of all processes. Thus, each actual entity is a process in which the experiences it has with other processual subjects merge together to form an integrated experience:

> "The final facts are, all alike, actual entities; and these actual entities are drops of experience, complex and interdependent" (ibid. 18).

Every process has experience-relations to other already existing processes that occupy concrete positions in space-time. It is these relations which

---

[29] Descartes, *Principles of Philosophy*, Part I, §51.
[30] Aristotle, *Categories*, Chapter 5.

make up the essence of the process. These kinds of relations, which are indispensable to the essence of the related entities, are usually called "internal relations". Whitehead calls them *prehensions*.

The third fundamental of Whitehead's ontology – the internal relationality *between* the actual entities – is an automatic consequence of the connection between processuality and subjectivity. If a process only comes about through experiencing relations with other processes, it cannot be disconnected even slightly from these relations, meaning that these relations are also necessarily internal.

It is arguably this inseparable connection between processuality and internal relationship that also creates the biggest difficulty in Whitehead's ontology: The process which gives rise to relations of experience exists prior to them neither logically nor temporally. The processual subject only comes into being through its relations with other subjects. This is something that can only be grasped intuitively, if at all, as it requires overcoming the boundaries of language, which demands that a subject exists before its predicates.

3.2. Process as growing together

It is against the background of these internal relations that the real distinctiveness of Whitehead's concept of process becomes apparent:

> "This internal relatedness is the reason why an event can be found only just *where* it is and *how* it is, – that is to say, in just one definite set of relationships. [...] This is what is meant by the very notion of internal relations" (1953, 155; italics by S.K.).

The essence of a primary entity, *what* it is, is inseparably connected to the place *where* it is (Whitehead 1979, 59f.) – "[t]hus an actual entity never moves: it is *where it is* and *what it is*" (ibid. 73; italics by S.K.). Both quotations also highlight that an actual entity (or actual occasion) cannot change ("how it is", "what it is"). In sharp contrast to the idea of substance, and especially its essential assumption of continuously persisting entities which can be integrated into various material structures,[31] Whitehead's processes "cannot have any external adventures", but "only the internal adventure of becoming. Its birth is its end" (ibid. 80). Thus, Whitehead's idea of process consists in a single act of becoming that elapses almost immediately after its completion. This understanding of process has little in common with the day-to-day notion of process, which also includes movement and change.

The core idea of Whitehead's metaphysics of process is that the self-creation of a process is the growing together of the many already completed (but not yet elapsed) prehended entities to form a new actual entity. After their completion, these actual occasions exist in space for only a short time and can be prehended. Whitehead calls the process, i.e., the new actual entity that arises from the integration of the prehensions *concrescence*, from the Latin verb "concresco" meaning "growing together". Therefore, the actual entities are acts of self-constitution, processes of forming a certain configuration. The essence of an actual entity includes *experiencing* its own self-creation to a new unity by means of the integration of its internal relations.

3.2.a. A different conception of causality – self and Umwelt

The growing together of the prehensions of a process to a unity of experience cannot be a deterministic procedure. For this to be the case, the process would have to be controlled by factors which influence it in a predetermined, i.e. in a non-processual, way, that is, by leading it to a predetermined end result. As non-processual factors they would have to be entities existing independently of the inner dynamics of the process which the process has incorporated through its prehensions. However, the nature of actual entities precludes the possibility of entities whose essence is unchangeably determined from controlling their development. This is because Whitehead does not conceive of "concrescence" as the recombination of prehended contents to a new conglomerate as if they were atomic modules.

[31] For example, Democritus and the mechanistic Atomists of the modern era considered atoms to be everlasting particles moving in a vacuum.

"Growing together" means that the integrated contents are broken down into their elements, which are then synthesized into a new whole. The process is controlled neither by deterministic efficient causation nor by unchangeable teleological final causation. The processual essence of actual entities consists in a self-creation for which the rules and facts of ideal and physical reality (i.e., maths, logic, mental content, the laws of nature, physical facts etc.) only provide the general framework of possibilities without determining the form of its self-creation. Therefore, every process contains a non-reducible *spontaneity* which results in its autonomous determination of its essence being unpredictable for ontological and not just epistemological reasons.

The idea of concrescence reveals an understanding of causality unique in the history of philosophy. The singularity of Whitehead's conception of causality consists in the fact that only things allowed into a process through its prehensions – meaning, ultimately, by the process itself – have causal relevance to this process. In other words: nothing external to an actual entity determines it – not even God, who Whitehead conceives of as being the most comprehensive process that coordinates all other processes. The factor which determines what can become an efficient cause for a given process is the subjectivity of the process itself. The process is a "teleological self-creation" (Whitehead 1967, 195), an act that creates its own teleology. It is teleological, not in the sense of substance of old metaphysics (which strives towards the aim determined by its fixed essence), but rather in the sense of a *processual teleology*: The process strives to determine its own essence. Finding its aim means determining the physical form which the completed process will have as a spatio-temporal fact. This striving towards finding its own aim is experienced by the process. The experience develops out of the evaluation of the relevance of prehended content for the process itself. Therefore, it is the teleology (or final causality) of the processual subject which decides what part of its physical surroundings can become an efficient cause, what can be integrated as a causal factor into the process and how this integration will occur.

Each process of concrescence necessarily implies a distinction between the facts of its physical surroundings which are allowed to be integrated in the process and those which have been negatively selected. Thus each process of conscrescence, even the most primitive one, exhibits an essential similarity to living beings: it is a self and at the same time, necessarily, defines its *Umwelt.*

The logic of this conception of causality is practically diametrically opposed to the logic of classical physics, in which efficient causes determine the course of an event.[32] This means that the prehended facts of the past do not "push" the process into the future in the way in which the causality of classical physics does (including the theories of relativity, thermodynamics and dynamic systems theory). With Whitehead's idea of process, the present of a process has a special meaning: The more complex a process is, the less of what is happening in the present is a simple continuation of the past. The present is not the passive and trivial transition from a completed past into a predetermined future. This is because the process decides, in its present, which factors from the past are to be considered relevant and which role the selected factors will have in forming the future. It is also because of this creative decision-making process that an actual occasion persists for a certain amount of time, as a creative act cannot be infinitely short like the infinitesimal time interval dt in physics. To put it simply, creativity takes time.[33]

3.2.b. Meta-physical "movement": a jump into space-time

The many completed processes that a developing process encounters and prehends open up various possibilities of combining the prehended multiplicity into a new whole for the new concrescence. The process of concrescence comes to its completion when all indeterminacies in relation to the realization of possibilities are eliminated. It is then that the new actual entity appears as a spatio-temporal fact. This concluding manifestation as spatio-temporally localized energy quantum is the expression of the act of a *decision* that consists in the realization of a single possibility for growing together (concrescence) and thereby the rejection of the remainder. In this way, the process internalizes a part of reality that is inseparably connected

[32] Even in chaos theory the possible courses of development of a non-linear dynamics are determined.

[33] This is also the core idea of the process philosophy of Henri Bergson.

to it and gives it new meaning by creating and manifesting itself as a new part of reality. Whitehead's process is an elementary vibration, a "polar occurrence of internalization and externalization" (Wiehl 1990, 228f.), whose birth as a spatio-temporal datum is only possible at the conclusion of its becoming, i.e., on determination of its own essence. It contains an "inner ambition which seeks to realize a particular energy quantum in a particular spatio-temporal region of the extensive continuum" (ibid.). In this way, Whitehead's process is not a physical, but rather a meta-physical "movement": "a movement before the movement of movable things" (Wiehl 1991, 326). It is a "jump" from a non-physical realm of being into physical reality (Whitehead 1953, 45). It is with this jump that, at the end of its becoming, the actual occasion achieves spatio-temporal existence and can be prehended by other processes.

Many modern interpretations of Whitehead's process philosophy agree that the concept of the actual entity is an inspiring description of *quantum-physical* actualization processes.[34] This involves the collapse of wave functions to spatio-temporally localised particles – in other words, to micro-physical entities which only manifest themselves spatio-temporally for an infinitesimally short period of time at the end of this collapse.

3.3. Processes form societies

A characteristic common to all actual entities – with the exception of God – is that they do not persist for long periods of time. As processes their justification for existence is lost as soon as they have completed their becoming or, in other words, when they have a defined essence like the substances of classical metaphysics.

Only entities that consist of groups of actual occasions are able to have a longer lifespan. Whitehead uses the term *nexus* to collectively describe all possible forms of accumulation of interconnected processes. Nexūs[35] are groups whose members form a connection – which can be either loose

---

[34] Chew 2004, 88; Stapp (this volume), 2004, 100; Malin 2004, 80; Jungerman 2000, 83; Lowe 1990, 232; Sherburne 1961, 78f.; Fetz 1981, 252.

[35] Whitehead uses "nexūs" as the plural of "nexus".

or compact – by means of prehensions. Whitehead considers objects of our sensory and scientific experience whose parts are connected in a complex and coherent way (e.g., cliffs, organisms, ecosystems, molecules, atoms, elementary particles, computers, buildings, planets, galaxies, vortexes, flames, and so on) to be a special form of nexūs which he calls *societies*. All societies have a common element of form which persists in time.

It is this element of form, which Whitehead also calls "defining characteristic" (1979, 34), which defines the form of a society. It allows a society to move coherently and to constantly develop. This element preserves the identity of the society and is inherited by the members of the society (the actual occasions) from their predecessors. Nearly all societies consist of parallel "strands" of inheritance and passing-down by means of prehensions (ibid. 35). In this case "strand" means a sequence of actual occasions in which each process is connected (via prehensions) more closely to one particular other process in its immediate past that to any other process. However, these strands are interwoven like the strands of a fabric, because the members prehend not only their immediate predecessor (which they replace) but also the members of neighboring strands, albeit much less intensively.

In this way, the individual processes that make up a society are knots in a network of mutual interdependence of essence. This network of internal relations gives the societies the stability necessary for a continuous development in space-time. Because of their internal relations, societies are markedly different from the network of efficient causation described by the dynamic systems theory of physics. The elements of such networks are thought of as entities that have a fixed essence (e.g., molecules or genes) that allow only external relations between them.

3.4. Abstract entities

The "defining characteristics" are forms that can be abstracted from the physical configuration of societies. These allow us to compare different societies to one another and to classify them. Abstract forms are complex conceptual entities that can be analyzed in simpler forms, such as numbers, geometric figures or types of elementary physical entities (e.g., electrons or photons). It is this that makes the scientific examination of nature possible.

According to Whitehead, abstract forms are not only of great importance but also have their own existence. He considers them to be timeless abstract entities which he calls *eternal objects*. They are "pure potentials for the specific determination of fact, *or* forms of definiteness" (ibid. 22). In process philosophy they assume the function exercised by ideas in Plato's metaphysics and the universals of scholasticism (ibid. 44; 1958, 32). However, they differ from these in ways that cannot be discussed here. Eternal objects can be seen as universal forms which, in contrast to the forms in Aristotelian metaphysics (eide), are not actively forming factors.

In conceiving of eternal objects in this way, Whitehead's metaphysics exhibits a close connection to the philosophy of Plato. Whitehead himself considers his own philosophy to be a contemporary renewal of Plato's thought (1979, 39). Plato's influence on the metaphysics of process philosophy can be clearly seen in the adoption of such central Platonic terms as "participation". For example, the purpose of actual entities requires the participation in eternal objects:

> "The things which are temporal arise by their participation in the things which are eternal" (ibid. 40).

An actual entity that is in the process of self-configuration prehends not only other actual entities but also eternal objects. The latter bring ideal forms of being into the developing process. They show the emerging process the clearly defined possibilities available to it from which it must choose.

3.5. Living societies, living occasions, and the entirely living nexus – the origin of organismic self and Umwelt

Living creatures are a special kind of society. Central to Whitehead's understanding of life is that a society can only be regarded as being alive if it also includes actual occasions whose mental pole is of considerable originality (ibid. 103). Whitehead calls these processes *living occasions* (ibid.

104). The totality of living occasions of a living creature amounts to an "entirely living nexus". A society is only then alive if it is controlled by such a nexus (ibid. 103). The entirely living nexus is the core of a *living society* (Whitehead's term for living beings) without being a society itself (ibid.). This is the case because its members are far too creative to limit themselves to the inheritance and passing-on of a particular defining characteristic. Living occasions add something to the life of an organism that had not been realized previously in either the entirely living nexus or the remaining living society.

The entirely living nexus is markedly different from Aristotle's concept of the soul. It is thought of in terms of process-metaphysics and, moreover, is not ontologically different from the rest of the living society – it is composed of actual occasions in the same way that the rest of the living creature is. In contrast to Aristotle's biophilosophy, which differentiates between an active and passive principle (i.e., between the soul and the body), there is no *ontological dualism* between living occasions and the rest of the living society. Therefore, Whitehead is no "cryptovitalist" as claimed by Ernst Mayr (2000, 353), since he does not assume that, living beings are equipped with a specific vital force alien to non-living matter, which Mayr considers to be the essence of vitalism (ibid. 418). Living occasions are nothing but actual entities. Their distinctiveness consists in their spontaneity, as their development does not conform entirely to the past which they inherit. For this reason, they cannot be explained only by the laws of nature. They bring something new to the history of a living creature that cannot be explained with the conceptions of causality of natural science, because it is not only the effect of the past of the living creature and its physical surroundings. Living occasions exhibit a stronger *selfhood* than ordinary processes of concrescence. As a result, in living occasions the distinction between *Umwelt* and mere physical surroundings is stronger than in ordinary actual occasions.

The foundation of any biophilosophy inspired by Whitehead must be the idea that Jonas, too, identified as a basic principle of his biophilosophy – "that the organic even in its lowest forms prefigures mind" (2001, 1).[36] From the perspective of a Whiteheadian biophilosophy, bio-

---

[36] In this instance Jonas was clearly influenced by Whitehead, who claimed that *mental* factors play a decisive causal role in *every* organism, even the most simple. However,

mathematics would only be able to explain living creatures if they had no living occasions, that is if they were only societies. Any kind of thinking that is based on efficient causation (it is this kind of thinking that forms the basis of biomathematical modeling) requires that certain regularities, be they universal (the laws of nature) or only valid for a particular organism (emergent laws), exert permanent influence on an organism throughout its whole life-cycle. Whiteheadians recognize this point but do not see this as the foundation of life. From the point of view of a Whiteheadian bio-philosophy, only the material manifestation of the living (i.e., the final manifestation of its processuality in space-time) can be adequately represented by using mathematical formulas; the creative base of livingness is outside the realm of quantitative-metric analysis.

Whiteheadian biophilosophy attempts to understand life's essence on the basis of a new metaphysics that is not substance-ontological or physicalistic as in the metaphysics of contemporary biosciences and philosophy of biology. Besides, Whiteheadian biophilosophy differs also from post-modern biophilosophy, because it does not emphasize the incomprehensibility of life which assumes life to have no essence.[37]

4. This book

All authors of the present anthology explicitly or implicitly focus on one or more of the four essential aspects of life mentioned above, as they are necessarily essential aspects of a Whiteheadian biophilosophy as well. The chapters of the book can be subdivided into three main thematic units: theory of organism, quantum biology, and evolution theory.

Barbara Muraca addresses the question of organismic teleology within biosciences from the point of view of both Kant's critical philosophy and Whitehead's ontology. She criticizes the anti-teleological intellectual attitude of most contemporary bioscientists because the current discussions about theories of self-organization and complexity and their applications

unlike Jonas, Whitehead does not make a strict separation between living and non-living things, as he supposes (proto)mental factors to exist in all quantum-physical processes.

[37] See Koutroufinis in this book (sections 1.5, 2.2, 2.3)

within biology and ecology give new significance to the idea of teleology. Starting her analysis by distinguishing amongst different concepts of teleology, Muraca emphasizes the role of "internal purposiveness" in today's biology. She shows that purposiveness corresponds to a complex form of reciprocal causation. On the basis of Kant's analysis of "natural purposes" in the Critique of Judgment as well as her own criticism of self-organization theory Muraca argues that reciprocal causation is not sufficient to describe organisms adequately. She claims that a genuine teleology of nature implies the idea of anticipation. Finally, Muraca shows that Whitehead's "philosophy of organism" provides the ontological framework for a theory of organismic anticipation by avoiding any recourse to supernatural forces.

Gernot and Renate Falkner focus on the adaptive response of a unicellular organism to alterations of nutrition supply and discuss possible analogies between the process of physiological adaptation and Whitehead's "actual occasion" of experience. They show that physiological adaptation is based on a sequence of adaptive experiential events. The authors postulate that in each adaptive event, initiated by an environmental (*Umwelt*) alteration that perturbs a previously attained adapted state, organisms experience a state of tension when energy converting subsystems are not optimally conformed to each other. Energy converting subsystems of the cell conform to each other, until a new adapted state emerges. This process of physiological adaptation is composed of individual adaptive events that have a bipolar nature: in an initial phase the changing external concentration, perceived with the cellular constituents resulting from former adaptations, is interpreted with respect to an appropriate reconstruction of the cell for future activities. The result of this anticipatory interpretation then leads in the final phase to a new cellular constituent, whose manifestation is then interpreted in subsequent adaptive events. In this regard adaptive events share essential features of Whitehead's "acts of becoming", by which an organismic self constantly re-creates itself in an experience of environmental (*Umwelt*) changes.

Spyridon Koutroufinis aims to demonstrate the suitability of some of Whitehead's main ideas for a natural philosophy of organismic ontogenesis based on a process-metaphysical understanding of teleology. He criticizes the assumption from which most bioscientists have proceeded, that organisms arise and preserve themselves by means of efficient causation and that only blind forces such as those studied by physics and chemistry, are at work in organisms. In the first part of his paper Koutroufinis shows that thinking of embryogenesis only in terms of efficient causation, which operates on the basis of the theory of nonlinear dynamical systems, poses serious problems. However, from the perspective of dynamical systems theory, which at the present time dominates theoretical biology, it makes sense to assume that each organism, during its own ontogenesis, often faces different possibilities of further development. On the basis of this assumption, in the second part of the paper it is argued that the Whiteheadian conceptions of the "actual entity" and the "entirely living nexus" allow one to consider ontogenetical developments as results of protomental teleological decisions between different possibilities of further development, without falling back to a vitalistic position or violating physical laws.

Jonathan Delafield-Butt proceeds on the assumption that purposeful behaviors of organisms fundamentally require prospective control to anticipate the future present. He presents two separate streams of thought that are closely analogous. The first is the process-metaphysics of Whitehead's "actual occasion" and the second is a perceptuomotor control theory from ecological psychology based on the "general tau theory". Both of these approaches explain a process of sensing, integrating, and acting in the world, but where the latter explains this process as occurring through space-time in a living animal, the former considers the process as a fundamental ontological construct. The juxtaposition of the two helps to inform each theory and so broaden our understanding of the component elements of the ontological unit and the psychophysical construct of a perception-action cycle. There are fundamental similarities between Whitehead's ontological unit and the unit of action described by general tau theory, since they are both teleological psychophysical units.

Joseph Earley focuses on Whitehead's conviction that indeterminacy is essential for both life and mind and on his assumption that "life is a characteristic of 'empty space'". Earley suggests that Whitehead's "'empty' space" should be considered as a metaphorical space of indeterminacy and claims that necessary indeterminacy is likely to emerge in networks of inner-organismic relationships which are considered by systems biology. By reference to bistable chemical dynamics Earley demonstrates how a complex dynamic system can give rise to indeterministic behavior. Finally, he sketches a Neo-Whiteheadian metaphysical approach called "Process Structural Realism" that can incorporate both the findings and the essential ideas of systems biology.

Pete Gunter focuses on the recent emergence of quantum decoherence theory with the notion of entanglement and its rejection of the human observer as a necessary component of measurement. These developments, along with the discovery of quantum effects occurring at the level of large molecules and large collections of atoms (mesoscopic quantum effects), and quantum nonlocality now make it possible to reconceive organismic dynamics in non corpuscular-kinetic terms. Gunter explores attempts to create quantum biologies of the organism by Johnjoe McFadden, Mae-Wan Ho, and Peter Gariev. He concludes that in several respects quantum biology is markedly congenial with Whitehead's philosophy of nature. However, some reworking of the Whiteheadian metaphysics seems to be required in order to make it applicable to the new quantum theory.

Henry Stapp tries to integrate the role of human experience in thinking about mind-brain relationships. Based on quantum physics he provides a theory about how our conscious thoughts can affect our physically described brains. This theory depends on the shift from the mechanical conception of nature to the psychophysical conception that emerged from the findings of the pioneers of quantum theory. According to Stapp this shift converted the role of our conscious thoughts from that of passive observers of a causally closed physically described universe to that of active participants in an essentially psychophysical understanding of nature. Stapp unfolds a theory about the mind-brain relationship starting from the assumption that psychophysical quantum events which can be described as Whiteheadian actual occasions take place in the brain.

John Cobb criticizes the standard Neo-Darwinistic explanation of evolution as only dealing with the random mutation of genes, the organisms that result from these, and the environment as selective agent. This enables biologists to think that a "materialist" account, one that excludes such things as the purposive behavior of animals based on their experiences, is adequate. However, Cobb argues, there is evidence that the actual course of evolution is far more complex. Far from being passive recipients of the effects of genetic change and environmental selection, organisms actively participate in shaping the course of evolution. Their activities affect the selection of genes through the Baldwin effect and also the character of the environment that selects. Cobb claims that the activity of organisms should be considered as an independent variable in any explanation of evolution. He thinks that resistance to accepting this comes largely from the desire of biologists to exclude any reference to the subjective, experiential side of organisms from their explanations. If they insist on doing so, they should recognize that their explanations are incomplete. According to Cobb the alternative is to expand the understanding of science to include the testing of hypotheses about the subjective side of nature.

Andrew Packard tries to deflect Cobb's criticism of the simplified version of Neo-Darwinism by providing three main arguments. First, that most of those who "accept" Neo-Darwinistic evolution theory are not required to test it and that its place in current teaching reflects cultural expectations and realities. Second, that all Neo-Darwinian formulations of evolution that are about the fate of genes or populations have a future reference: uncoupled, therefore, from the work of most biologists concerned with living processes in the present – or the story of evolution in the past. Third, there is a long tradition of Darwinian biologists who include subjective aspects of the organisms they study and ascribe to behaviour and the activities of the phenotype an important role in directing the course of evolution. In the second half of the chapter Packard approaches Whiteheadian understanding of the role of subjectivity by drawing on his own experience of psycho-physics and of the forms taken by pattern-recognition in the life histories of aquatic organisms.

David Griffin shows that what is generally considered the Neo-Darwinian theory of evolution can be characterized in terms of 13 doctrines, some empirical and some metaphysical in nature. Much of the discussion of Neo-Darwinism by both advocates and detractors is confused because it is not clear which of the 13 doctrines the speaker has in mind. Griffin suggests that from a Whiteheadian point of view, 4 of these doctrines are true, the other 9 false. Most of the false doctrines are ones that imply an atheistic worldview. Griffin argues that the atheism of Neo-Darwinism has led it to be strongly opposed by those who advocate "creation science" or at least "Intelligent Design". But these views are also unsatisfactory from a Whiteheadian point of view, because they reject the doctrine of naturalism, which is one of Neo-Darwinism's true doctrines. Griffin refers to naturalism in the generic sense, which simply rules out supernatural interruptions of the world's normal cause-effect relations (not the sensationist-atheist-materialist version of naturalism). After showing why the advocates of Neo-Darwinism and Intelligent Design usually talk past each other and why neither can see the elements of truth in the other view, Griffin suggests a Whiteheadian theory of evolution that, being naturalistic but not atheistic, and nondualistic but not materialistic, unlike Neo-Darwinism, is acceptable from a religious-moral perspective and one that, unlike Intelligent Design, is acceptable from a scientific point of view.

Robert Valenza claims that Whiteheadian metaphysics in particular and dual aspect theories in general allow that reality coheres in knots that admit experience. Such entities may carry a subject-centered phenomenal aspect, and some of the more complex ones also manifest a perspective-free epistemological aspect. Both aspects supply part of the basis for a rational ontology, but it is the latter that affords the possibility of a worldview, and, in particular, a community-wide ontological deployment that can be fully shared. Valenza claims that subjects distinguish world objects on the basis of a generalized conception of symmetry that often goes by the name covariance. In this light Valenza explores the hypothesis, suggested by the history of the world and science, that nature moves systematically toward the development of a covariant epistemology, and that this is reflected in the evolution of life forms of increasing complexity. Valenza claims that explanations of this dynamic might include, among others, Whitehead's theory of the relationship of God to actual entities in general.

REFERENCES

Ariew, André. "Teleology". In *The Cambridge Companion to the Philosophy of Biology*, edited by David L. Hull and Michael Ruse, 160-181. New York: Cambridge University Press, 2007.

Aristotle (1999). *Physics* (translated by R. Waterfield). Oxford: Oxford University Press.

—— (2001). *On the Soul* (translated by J. Sachs). Santa Fe, NM: Green Lion Press.

—— *Categories* (translated by E.M. Edghill). The Internet Classics Archive http://classics.mit.edu/Aristotle/categories.html

—— (1963). *Categories and De Interpretatione* (translated by J.L. Ackrill). Oxford: Oxford University Press.

Ashby, W. (1962). "Principles of Self-Organizing Systems". In: Foerster, H. v.; Zopf, G. Jr. (eds.). *Principles of Self-Organization*. London: Pergamon Press, pp. 255-278.

—— (1947). "Principles of the self-organizing dynamic system". In: *Journal of General Psychology*, 37, pp. 125-128.

Beckner, M. (1959). *The Biological Way of Thought*. New York: Columbia University Press.

Bedau, M. (1998). "Where is the Good in Teleology?" In: Allen, C; Bekoff, M; Lauder, G. (eds.) *Nature's Purposes. Analysis of Function and Design in Biology*. Cambridge, MA: MIT Press, pp. 261-291.

Bertalanffy, L. v. (1932). *Theoretische Biologie*. Berlin: Bornträger. (in German)

—— (1952). *Problems of Life: An Evaluation of Modern Biological and Scientific Thought*. New York: Harper.

Bohr, N. (1933). "Licht und Leben". In: *Naturwissenschaften* 21 (13), pp. 245-250. Also published in: Küppers, B.-O. (ed.) (1990). *Leben = Physik + Chemie?* Munich, Zurich: Piper, pp. 35-47. (in German)

Brandon, R. (1990). *Adaptation and Environment*. Princeton, NJ: Princeton University Press.

Chew, G. (2004) "A Historical Reality that Includes Big Bang, Free Will, and Elementary Particles". In: Eastman, T.; Keeton, H. (eds.). *Physics and Whitehead*. Albany: State University of New York Press, pp. 84-92.

Christensen, W. (1996). "A Complex Systems Theory of Teleology". *Biology and Philosophy* 11: pp. 301-320.

Christian, W. (1967). *An Interpretation of Whitehead's Metaphysics*. New Haven: Yale University Press.

Collini, E.; Wong, C.; Wilk, K. et al. (2010). "Coherently wired light-harvesting in photosynthetic marine algae at ambient temperature". In: *Nature* 463, pp. 644-647.

Darwin, Ch. (1989). *The Descent of Man, and Selection in Relation to Sex*. In: Barrett, P.H.; Freeman, R.B. (ed.). *The works of Charles Darwin* (vol. 21), London: The Pickering masters.

Davia, C. (2006). "Life, Catalysis and Excitable Media: A Dynamic Systems Approach to Metabolism and Cognition". In: Tuszynski, J. (ed.). *The Emerging Physics of Consciousness*. Berlin, Heidelberg: Springer, pp. 254-292.

Deacon, T. (2012). *Incomplete Nature*. New York: W. W. Norton & Company.

—— (2006). "Reciprocal Linkage Between Self-Organizing Processes is Sufficient for Self-reproduction and Evolvability". In: *Biological Theory* 1, no. 2, pp. 136-149.

Deacon, T.; Koutroufinis, S. (forthcoming). "Information, complexity, and dynamic depth". In: *Information*.

De Caro, M.; Macarthur, D. (eds.) (2010a). *Naturalism and Normativity*. New York: Columbia University Press.

—— (2010b). "Introduction : Science, Naturalism, and the Problem of Normativity". In: De Caro, M.; Macarthur, D. (eds.) (2010a). *Naturalism and Normativity*. New York: Columbia University Press, pp. 1-19.

De Caro, M.; Voltolini, A. (2010). "Is Liberal Naturalism Possible ?" In: De Caro, M.; Macarthur, D. (eds.) (2010a). *Naturalism and Normativity*. New York: Columbia University Press, pp. 69-86.

Descartes, R. (1993). *Meditations on First Philosophy* (edited by S. Tweyman, translated by E.S. Haldane and G.R.T. Ross). London, New York: Routledge.

—— (1984). *Principles of Philosophy* (translated by V. Rodger). Dodrecht, Boston, Lancaster: Reidel.

Dupré. J. (2010). "How to be Naturalistic Without Beeing Simplistic in the Study of Human Nature". In: De Caro, M.; Macarthur, D. (ed.) (2010a). *Naturalism and Normativity*. New York: Columbia University Press, pp. 289-303.

—— (2004). "The Miracle of Monism". In: De Caro, M.; Macarthur, D. (eds.) (2004). *Naturalism in Question*. Cambridge, MA; London, England: Harvard University Press, pp. 22-39.

Elsasser, W. (1990). "Eine Kritik am Reduktionismus". In: Küppers, B.-O. (ed.). *Leben = Physik + Chemie?* Munich, Zurich: Piper, pp. 211-236. (in German)

—— (1982). "The Other Side of Molecular Biology". In: *Journal of theoretical Biology* 96, pp. 67-76.

Emmet, D. (1981). *Whitehead's Philosophy of Organism*. Westport, CT: Greenwood Press.

Engel, G.; Calhoun, T.; Read, E. (2007). "Evidence for wavelike energy transfer through quantum coherence in photosynthetic systems". In: *Nature* 446, pp. 782-786.

Fetz, R. (1981). *Whitehead: Prozeßdenken und Substanzmetaphysik*. Freiburg, Munich: Alber. (in German)

Foerster, H. v.; Zopf , George W. Jr. (eds.) (1962). *Principles of Self-Organization. The Illinois Symposium on Theory and Technology of Self- Organizing Systems*. New York: Pergamon Press.

—— (1960). "On Self-Organizing Systems and Their Environments". In: Yovits, M.; Cameron, S. (eds.). *Self-Organizing Systems*. London: Pergamon Press, pp. 31-50.

Glansdorff, P.; Prigogine, I. (1971). *Thermodynamic Theory of Structure, Stability and Fluctuations*. Chichester: Wiley-Interscience.

Goldbeter, A. (1996). *Biochemical Oscillations and Cellular Rhythms*. Cambridge, New York: Cambridge University Press.

Goodwin, B. (1994). *How the Leopard Changed Its Spots*. New York: C. Scribner's Sons.

Griffiths, P. (ed.) (1992). *Trees of Life*. Dodrecht, Boston, London: Kluwer.

Haken, H. (1983). *Synergetics: An Introduction*. Berlin, New York: Springer.

—— (1973). *Synergetics: cooperative phenomena in multi-component systems*. Stuttgart: Teubner.

Hameroff, S. (2007). "Orchestrated Reduction of Quantum Coherence in Brain Microtubules. A Model for Consciousness". In: *NeuroQuantology*, Vol. 5, Issue 1, pp. 1-8.

—— (2003). "Consciousness, Whitehead and Quantum Computation in the Brain: Panprotopsychism Meets the Physics of Fundamental Space-Time Geometry". In: Riffert, F.; Weber, M. (eds.). *Searching for New Contrasts*. Frankfurt/M.: Peter Lang, pp. 61-86.

Hameroff, S; Tuszynski, J. (2004). "Quantum states in proteins and protein assemblies: The essence of life?" In: Abbott, D.; Bezrukov, S.; et al. (eds.). *Fluctuations and Noise in Biological, Biophysical, and Biomedical Systems II*, Proceedings of SPIE – Vol. 5467, pp. 27-41.

Hampe, M. (1998). *Alfred North Whitehead*. Munich: Beck. (in German)

Heisenberg, W. (1984). "Das organische Leben". In: Blum, P.; Dürr, H.-P.; Rechenberg, H. (eds.). *Ordnung der Wirklichkeit*. Munich: Piper, pp. 259-273. Also published in: Küppers, B.-O. (ed.) (1990). *Leben = Physik + Chemie?* Munich, Zurich: Piper, pp. 49-72. (in German)

Heitler, W. (1976). "Über die Komplementarität von lebloser und lebender Materie". In: *Abhandlungen der Mathematisch-Naturwissenschaftlichen Klasse der Akademie der Wissenschaften und der Literatur in Mainz*, Nr. 1, pp. 3-21. Also published in: Küppers, B.-O. (ed.) (1990). *Leben = Physik + Chemie?* Munich, Zurich: Piper, pp. 189-210. (in German)

Hull, D. (1974). *Philosophy of Biological Science*. Englewood Cliffs, NJ: Prentice-Hall.

Hull, D.; Ruse, M. (ed.) (2007). *The Cambridge Companion to the Philosophy of Biology*. Cambridge, New York et al.: Cambridge University Press.

Jonas, H. (2001). *The Phenomenon of Life*. Evanston, IL: Northwestern University Press.

Jordan, P. (1972). "Über die exobiologische Hypothese". In: *Erkenntnis und Besinnung*. Oldenbug: Stalling, pp. 175-206. Also published in: Küppers, B.-O. (ed.) (1990). *Leben = Physik + Chemie?* Munich, Zurich: Piper, pp. 159-188. (in German)

—— (1947). *Die Physik und das Geheimnis des organischen Lebens*. Braunschweig: Vieweg. (in German)

Jungerman, J. (2000). *World in Process*. Albany: State University of New York Press.

Kauffman, S. (2008). *Reinventing the Sacred*. New York: Basic Books.

—— (2002). "What is Life?" In: Brockman, J. (ed.). *The next fifty years*. New York: Vintage Books, pp. 126-141.

—— (2000). *Investigations*. Oxford, New York: Oxford University Press.
—— (1995). *At Home in the Universe: The Search for Laws of Self-Organization and Complexity*. New York: Oxford University Press.
—— (1993). *The Origins of Order: Self-Organization and Selection in Evolution*. New York: Oxford University Press.
King, C. (2006). "Quantum Cosmology and the Hard Problem of the Conscious Brain". In: Tuszynski, J. (ed.). *The Emerging Physics of Consciousness*. Berlin, Heidelberg: Springer, pp. 407-456.
—— (1996) "Neurodynamics and Quantum Chaos: Resolving the mind-brain paradox through novel biophysics". In: Mac Cormac, E.; Stamenov, M. (eds.). *Fractals of brain, fractals of mind*. Amsterdam, Philadelphia: John Benjamins, pp. 179-233.
Koutroufinis, S. (2013). "Teleodynamics: A Neo-Naturalistic Conception of Organismic Teleology". In: Henning, B.; Scarfe, A. (eds.). *Beyond mechanism: Putting Life Back Into Biology*. Lanham (MD): Lexington Books/Rowman & Littlefield, pp. 309-342.
—— (1996). *Selbstorganisation ohne Selbst*. Berlin: Pharus-Verlag. (in German)
Koutroufinis, S.; Wessel, A. (2013). "Toward a Post-Physicalistic Concept of the Organism". In: *Annals of the History and Philosophy of Biology* 16, pp. 29-50.
Kraus, E. (1979). *The Metaphysics of Experience*. New York: Fordham University Press.
Lango, J. (1972). *Whitehead's Ontology*. Albany: State University of New York Press.
Leclerc, I. (1975). *Whitehead's Metaphysics*. Bloomington & London: Indiana University Press.
Lee, H.; Cheng, Y.; Fleming, G. (2007). "Coherence Dynamics in Photosynthesis: Protein Protection of Excitonic Coherence". In: *Science* 316 (no. 5830), pp. 1462-1465.
Leibniz, G. (1991). *Monadology* (translated by N. Rescher). Pittsburgh, PA: University of Pittsburgh Press.
—— *Principles of Nature and Grace*. http://www.earlymoderntexts.com/pdf/leibprin.pdf
Lotka, A. (1925): *Elements of Physical Biology*. Baltimore: Williams and Wilkins.
Lowe, V. (1990). *Alfred North Whitehead. The Man and his Work*, Vol. II. Baltimore, London: The John Hopkins University Press.
—— (1985). *Alfred North Whitehead. The Man and his Work*, Vol. I. Baltimore, London: The John Hopkins University Press.
—— (1966). *Understanding Whitehead*. Baltimore: The John Hopkins University Press.
Malin, S. (2004). "Whitehead's Philosophy and the Collapse of Quantum State". In: Eastman, T.; Keeton, H. (eds.). *Physics and Whitehead*. Albany: State University of New York Press, pp. 74-83.
Mayr, E. (2000). *Das ist Biologie*. Heidelberg, Berlin: Spektrum. (in German)

—— (1991). *Eine neue Philosophie der Biologie*, München, Zürich: Piper. (in German)
McDowell, J. (2004). "Naturalism in the Philosophy of Mind". In: De Caro, M.; Macarthur, D. (eds.) (2004). *Naturalism in Question*. Cambridge, MA; London, England: Harvard University Press, pp. 91-105.
Mershin, A.; Sanabria, H.; Miller, J. et al. (2006). "Towards Experimental Tests of Quantum Effects in Cytoskeleletal Proteins". In: Tuszynski, J. (ed.). *The Emerging Physics of Consciousness*. Berlin, Heidelberg: Springer, pp. 95-170.
Murray, J. (1993). *Mathematical Biology*. New York, Berlin, Heidelberg: Springer.
Neumann, J. v. (1966). *Theory of Self-Reproducing Automata*. Urbana, London: University of Illinois Press.
Nicolis, G.; Prigogine, I. (1977). *Self-organization in nonequilibrium systems*. New York: Wiley.
Noble, D. (2006). *The Music of Life*. Oxford, New York: Oxford University Press.
Portmann, A. (1960). *Die Tiergestalt*. Basel: Friedrich Reinhardt. (in German)
Prigogine, I.; Stengers, I. (1984). *Order out of chaos*. Toronto, New York: Bantam books.
Prigogine, I.; Lefever, R. (1968). "On symmetry-breaking instabilities in dissipative systems, II". In: *Journal for Chemical Physics* 48, pp. 1695-1700.
Prigogine, I.; Nicolis, G. (1967). "On symmetry-breaking instabilities in dissipative systems". In: *Journal for Chemical Physics* 46, pp. 3542-3550.
Rashevsky, N. (1938). *Mathematical biophysics: physico-mathematical foundations of biology*. Chicago: University of Chicago Press.
—— (1940). *Advances and applications of mathematical biology*. Chicago: University of Chicago Press.
Reinke, J. (1901). *Einleitung in der theoretischen Biologie*. Berlin: Verlag von Gebrüder Paetel. (in German)
Rosenberg, A.; McShea, D. W. (ed.) (2008). *Philosophy of Biology*. New York, London: Routledge.
Rosenberg, A. (1985). *The Structure of Biological Science*. Cambridge: Cambridge University Press.
Rosenblueth, A.; Wiener, N.; Bigelow, J. (1943). "Behavior, Purpose and Teleology". *Philosophy of Science* 10, no. 1, pp. 18-24.
Ruse, M. (1973). *The Philosophy of Biology*. London: Hutchinson & Co.
—— (1988). *Philosophy of Biology Today*, Albany: SUNY Press.
Rust, A. (1987). *Die organismische Kosmologie von Alfred N. Whitehead*. Frankfurt/Main: Athenäum. (in German)
Sayer, R. (1999). *Wert und Wirklichkeit*. Würzburg: Ergon. (in German)
Schaxel, J. (1919). *Grundzüge der Theorienbildung in der Biologie*. Jena: Gustav Fischer. (in German)
Schrödinger, E. (1944). *What is Life?* Cambridge: The University Press.

Sherburne, D. (1961). *A Whiteheadian Aesthetic*. New Haven: Yale University Press.
Stapp, H. (2004) "Whiteheadian Process and Quantum Theory". In: Eastman, T.; Keeton, H. (eds.). *Physics and Whitehead.* Albany: State University of New York Press, pp. 92-102.
Thacker, E. (2005). "Biophilosophy for the 21st Century". In: Kroker, A.; Kroker, M. (ed.). *1000 Days of Theory*. http://www.ctheory.net/articles.aspx?id=472#bio
Turing, A. (1952). "The Chemical Basis of Morphogenesis". In: *Philosophical Transactions of the Royal Society of London* (Series B, No.641, Vol. 237), pp. 37-72.
Uexküll, J. v. (1909). *Umwelt und Innenwelt der Tiere*. Berlin: Springer. (in German)
——(1920). 1920. *Theoretische Biologie*. Berlin: Verlag von Gebrüder Paetel. (in German)
Volterra, V. (1926). "Variazioni e fluttuazioni del numero d'individui in specie animali conviventi". In: *Memorie della R. Acc. dei Lincei* (Ser. VI, Vol. II), pp. 31-113. (in Italian)
——(1931). *Leçons sur la théorie mathématique de la lutte pour la vie*. Paris: Gauthier-Villars. (in French)
Whitehead, A. N. (1979) *Process and Reality*. New York: Free Press.
—— (1953). *Science and the Modern World.* Cambridge: At the University Press.
—— (1967). *Adventures of Ideas*. New York: Free Press.
—— (1958). *The Function of Reason*. Boston: Beacon Press.
Wiehl, R. (1991)."Aktualität und Extensivität in Whiteheads Kosmo-Psychologie". In: Hampe, M.; Maaßen, H. (eds.), *Die Gifford Lectures und ihre Deutung*. Frankfurt/Main: Suhrkamp, pp. 313-368. (in German)
——(1990). "Whiteheads Kant-Kritik und Kants Kritik am Panpsychismus". In: Holzhey, H.; Rust, A.; Wiehl, R. (ed.). *Natur, Subjektivität, Gott*. Frankfurt/Main: Suhrkamp, pp. 198-239. (in German)
Zahavi, A. (1975). "Mate selection—a selection for a handicap". In: Journal of Theoretical Biology 53(1): pp.: 205-214.
——(1997). *The Handicap Principle: a Missing Piece of Darwin's Puzzle*. Oxford: Oxford University Press.

Teleology and the Life Sciences: Between Limit Concept and Ontological Necessity

BARBARA MURACA

1. Introduction

Against the background of the current discussion about self-organization theories and complexity theories and their application within biology and ecology, the question of teleology gains a new significance. Some scholars insist on the total elimination of any reference to teleology from the realm of the natural sciences. However, it seems especially hard to eradicate teleological expressions from scientific language when the issue of understanding living beings is at stake. For this reason, other scholars opt for a middle path that allows for some teleological language. Yet, it is an open question whether teleological expressions are to be considered as playing a merely metaphorical or a necessary heuristic role in the sciences. Moreover, the ontological presuppositions, which underpin different positions in the debate, need to be depicted and analyzed.

This paper aims at addressing the question of teleology within the life sciences by taking into account both Kant's critical philosophy and Whitehead's ontology. My analysis starts with Georg Toepfer's distinction among different concepts of teleology and then focuses on the role of "internal purposiveness" (innere Zweckmäßigkeit) for biology today. I show how purposiveness (Zweckmäßigkeit; hereafter: ZM) corresponds to a very complex form of reciprocal causation (Wechselwirkung) rather than to any model of final causation. Drawing on Kant's analysis of "natural purposes" in the *Critique of Judgment* (CJ) as well as self-organization theory, I claim that reciprocal causation – however complex it might well be – is not sufficient to describe living beings adequately. However, since the natural sciences are still caught up in the presuppositions of modern scientistic and materialistic ontology, a step beyond mere efficient causation seems to be impossible within their methodological framework. And yet, as I will show, a genuine teleology of nature implies the idea of anticipation of totality. This kind of teleological consideration is presented at first in its role as a regulative concept in Kantian terms.

Finally, I follow the path of Whitehead's 'philosophy of organism' and claim for natural teleology the state of a necessary ontological presupposition. Whitehead's ontology offers an ontological underpinning for teleological issues that, by avoiding any recourse to supernatural forces, invites life and natural sciences to a fruitful dialogue at the limit of their methodological boundaries, pressing them beyond their unreflected presuppositions.

2. Teleology concepts

The teleology debate looks like an open field of discussion that hosts several players from different areas, all of them seeking a satisfactory definition and delimitation of the concept. As an orienting path in this field I follow Toepfer's detailed analysis of teleological concepts and consider them in their relevance to the life sciences.

By referring to Kant's *Critique of Judgment* Toepfer distinguishes between different meanings of teleology: special versus universal teleology; internal versus external teleology; purposiveness (ZM) versus intentionality (Zwecksetzung, hereafter: ZS).

2.1. Special versus universal teleology

Toepfer defines universal teleology as the idea of there being one general "orientation of the universe towards one single end" (Toepfer 2005, 36; own translation), which is present all the way through and leads everything towards that very end. The idea of universal teleology, however, does not necessarily imply the concept of God, although this is very often the case, as we know from Kant's critique of physiotheological theories (CJ § 85, 330ff). Universal teleology as such encompasses the idea of an orientation of nature as a whole towards a specific general end. However, since "the whole of nature" cannot as such be the object of any empirical approach, such an idea is by no means reconcilable with the models of explanations conceived by science. Therefore, the question about a general end of nature – no matter whether it is linked or not with the concept of God – is inevitably situated outside the methodological domain of biology and ecology and cannot be addressed from their point of view.

Nonetheless one might claim that the idea of a universal teleology could be a helpful instrument for science if applied as a regulative concept in a Kantian sense of the term. In this case, it would be necessary to make clear whether universal teleology offers to scientific research any added value in heuristic terms or whether it belongs merely to the private sphere of beliefs every researcher might or not entertain. No matter how the answer to this question turns out, we can at least hold at this point that even a complete rejection of universal teleology does not necessarily lead to a rejection of any form of addressing of teleological issues within sciences.

Beside universal teleology we can in fact recognize a different form of teleological reasoning called by Toepfer special teleology, which refers to single entities and does not imply a general aim for the whole nature. According to Toepfer, special teleology "attributes purposiveness (ZM) to single natural bodies (just as it is accomplished within biology)" (Toepfer 2005, 36; own translation). Biology and ecology, especially when they refer to organisms and their functioning, can hardly avoid teleological terms even if they claim a mere metaphorical employment of them.

Consequently, when we speak of teleology for the life sciences – be it in a heuristic or a metaphorical way – what is meant can only be special teleology.

2.2. Internal versus external teleology

While internal teleology according to Toepfer encompasses the inner teleological interdependence between parts within the whole of a single organism, external teleology refers to the chain of usefulness that something as a whole might have for something else (Toepfer 2005, 36), i.e. the chain linking means to ends. Through this distinction, which goes back to Kant's *Critique of Judgment*, we are faced with a concept of end merely associated with the external usefulness of services. More accurately, this concept does not only cover possible utilities for human beings but also in Kant's own words "the benefit for any other creature" (CJ § 63, 245). The teleological aspect, rather than being centered in a particular entity, concerns its relations to other entities, which might employ it as a means for the realization of their own purposes. There is no reciprocal exclusion between internal and external teleology, since even an entity, which is an end in itself, can be employed as a means for the realization of purposes external to it. In fact, this happens among human beings all the time: the orientation to external ends does not automatically inhibit the internal one. Ethical problems arise according to Kant only when somebody is reduced to a mere means, because in this case her inner teleology is neglected or even denied.

As McLaughin writes, we attribute functions to something when we consider it as a means to an end (McLaughin 2005, 19). By redefining ends in terms of functions, services, utility or benefit, external teleology turns into a familiar language for biology and allows scientists strategically to elude the thorny discussion about internal teleology.

However, it is questionable whether one can still speak about teleology at all without any reference to the issue of internal teleology. In fact, if we follow Kant's reasoning, we can truly speak about means and ends only if we acknowledge intentionality (ZS) – the internal positing of an end for which other creatures or objects work as a means. As Kant clearly asserted, "we can easily see from this that extrinsic purposiveness (a thing's being beneficial to others) can be regarded as an extrinsic natural purpose only under the condition that the existence of what it benefits proximately or remotely is a purpose of nature in its own right" (CJ § 63, 246).

In other words, it makes sense to talk about a means only if we acknowledge that there are some ends in themselves, to which it can be a means. Therefore, external teleology is intimately dependent on internal teleology for its very definition[1]. If we generally exclude any reference to ends in themselves the means-ends-chain can be easily reduced to a relation of efficient causation, in which the means represents no more than one out of many efficacious factors affecting the "end". As McLaughin writes:

[1] Since we cannot know about ends in themselves in nature, Kant speaks of external teleology with respect to human beings and the freedom of their causality. Thus every form of external teleology is grounded in human beings, who represent the end in itself, which is located at the bottom of the ends-means chain.

"In every material system we can hold each efficient cause of an effect as a means to this very effect, if we consider in turn the effect as an end" (McLaughin 2005, 19). In the case of inorganic entities – that is, when ends in themselves can be kept out of the picture – even Kant describes the means as the "closest cause" (CJ § 63, 240). With respect to nowadays well known complexity of causal relation, as it manifests itself eminently in non-linear processes, we might want to reformulate Kant's own expression in terms of "the most relevant cause". Anyway the relation between means and ends in the absence of internal teleology, which encompasses ends in themselves, falls back onto mere efficient causation: external teleology alone is no teleology at all.

2.3. Purposiveness, agency, and intentionality

Given the unavoidability of internal teleology for any sort of teleology-like language some scholars aim at developing a kind of internal teleology, which does not fundamentally question the methodological approach of the life sciences. Toepfer suggests thus a further distinction between purposiveness ZM and agency (Zwecktätigkeit, hereafter: ZT) or intentionality (ZS)[2]: while purposiveness (ZM) is not linked with will or mind in any form, the other two terms imply a "*mental* anticipation of future states" (Toepfer 2005, 37; italics BM) and can be ascribed exclusively to entities that can perform such an anticipation. Toepfer explains it as follows: we can properly think final causation in a strict sense only by admitting the idea of a back-causation from the future – one in which the end is "affecting/causing back". How can we conceive such a back-affection without falling into the paradox of a reversal in the direction of time? One feasible way for Toepfer is to consider it as an intentional setting of aims, in which the aims act back by means of anticipation. In other words it is the anticipation of the aim that leads the action and affects it in the form of final causation. Besides, since Toepfer equates anticipation with mind process-

---

[2] Toepfer does not distinguish between intentionality – Zwecksetzung, which is the (conscious) setting of an end – and agency – Zwecktätigkeit, which does not necessarily imply consciousness, as will be shown later.

es, he speaks of a symbolic anticipation tout court and considers language as a necessary element in it (ibid. 37ff).

The distinction between ends which are not in the least linked to intentionality and belong thus to the realm of purposiveness (ZM), and those which are set by a freely acting subject by means of anticipation, plays a fundamental role for the employment of teleological expressions within the sciences. Depending on how this distinction is intended, teleology can be included or not in contemporary scientific research. Some scholars like Mahner and Bunge endorse a definitive and total eradication of any reference to teleological language from biology since they consider it simply "wrong" (Mahner/Bunge 2000, 327, 376). Since according to them all teleological terms in the end draw on intentionality, it seems a hopeless venture to talk about teleology without taking for granted an aim-setting will. Moving a step further Stegmüller regards any use of teleological concepts, which does not include will or intellect as a semantic nonsense (Stegmüller 1996/83, 756). Such an assertion is rooted in the assumption that intellect and free will, understood as a peculiar human property, is a necessary presupposition for the anticipation that (according to Toepfer) is the condition for teleology as intentionality. However, there is no reason to reject the hypothesis that different forms of anticipation are possible, ones which do not necessarily depend on human intellect and free will. Many animals for example are quite capable of anticipation at the emotional level. If anything, it seems that intellect itself can be understood as the outcome of the higher development of that very emotional anticipation.

Since biology in spite of all critique seems to cling anyway to that very semantic nonsense in such a way that a complete dropping of teleological language is ultimately unavoidable, a teleological concept is needed that does not automatically imply human free will.

For this reason, Toepfer's concept of a special internal purposiveness (ZM) seems at first to meet this requirement[3]. Toepfer depicts purposive-

[3] The concept of teleonomy that Pittendringh introduced into biology in 1958 and Mayr further developed is very close to the concept of internal purposiveness (ZM) (Pittendrigh 1958). However, the term "teleonomy" as such does not offer any better solution to the debate, unless it is linked to a profound and detailed explanation of which specific understanding of teleology it is supposed to cover. In other words, the

ness (ZM) as a necessary and constitutive concept for the description of organisms and living beings. According to Toepfer, living beings are characterized by a peculiar form of causality, which he describes as the reciprocal relation between the different parts of a system (Toepfer 2005, 45). In this definition he clearly refers to Kant:

> "In such a product of nature, just as each part exists only as a result of all the rest, so we also think of each part as existing for the sake of the others and of the whole, i.e. as an instrument (organ)" (CJ § 65, 253).

This particular form of teleology is depicted in terms of reciprocal action and retroaction. Rather than being rectilinear, as would be the case for intentionality, it bears a cyclical structure (Toepfer 2005, 47). Toepfer elucidates the constitutive character of purposiveness (ZM) as an epistemological priority in the attempt to understand living organisms: in a reciprocal relation among the different parts biological research considers the effect as the "Bestimmungsgrund" (*ratio definitionis*), yet not as the "real cause of the object in question" (ibid., 50). In other words, with respect to a reciprocal action the effect is envisaged as a priority and therefore considered as an aim. Such a teleological concept bears thus a heuristic function for sciences and allows for the analysis of living beings.

However, it is at least questionable whether in this case we are still faced with genuine teleology. In fact, even by following Toepfer in his epistemological use of the concept it remains doubtful whether the reciprocal action he is talking about requires any further reference to teleology. Rather, if purposiveness (ZM) is intended as mere reciprocal action it does not need any teleological concept at all in order to be of some use for science.

Moreover, if we abandon the idea of linear causality as it is endorsed by classical physics and consider contemporary self-organization theories, we have to acknowledge that phenomena of reciprocal action can be fully explained within the framework and methodology of natural sciences with no recourse to any teleological language (compare here Prigogine's research about chemical dissipative structures in Prigogine/Stengers 1981).

shift to a brand new term cannot substitute for an analysis of the very discussion I have presented here. Thus in this paper I will not make use of the term teleonomy.

Moreover, the concept of retroaction nowadays is an ordinary category in science as it appears for example in synergetics.

To sum up, the assertion that reciprocal action and retroaction are constitutive concepts for biological research does not in the least lead to the same conclusion for teleological issues.

However, the description of living organisms in terms of mere reciprocal action is not sufficient for an appropriate understanding of them. This is why a reference to teleology that goes beyond Toepfer's definition of purposiveness (ZM) and includes a dimension of agency (ZT) (if not even intentionality (ZS)) is an unavoidable condition for biological research.

3. Kant's natural teleology

3.1 A new form of causality

The acknowledgment that causality as reciprocal action does not challenge the methodological approach of the natural sciences goes back to Kant. Since reciprocal action is one of the categories of the intellect it corresponds as a transcendental condition of possibility to sense perception. In other words, reciprocal action can very well be an object of perception and science. If purposiveness (ZM) in the end can be explained as a peculiar form of reciprocal action among the parts of a whole there is no reason why Kant should have had any trouble including it among the categories of knowledge. Yet, Kant writes: "Strictly speaking, therefore, the organization of nature has nothing analogous to any causality known to us" (CJ § 65, 254) and some lines further:

> "But intrinsic natural perfection, as possessed by those things that are possible only as natural purposes and that are hence called organized beings, is not conceivable or explicable, on any analogy to any known physical ability, i.e. ability of nature, not even – since we too belong to nature in the broadest sense – on a precisely fitting analogy to human art" (ibid.).

According to Kant it's impossible for us to conceive natural purposiveness (ZM) precisely because we are confronted with a form of causality that is by no means compatible with our cognitive faculty.

In fact, the peculiarity of natural teleology does not primarily consist in the mutual affection of the parts; rather, it refers to the fact that "the parts of the thing combine into the unity of a whole because they are reciprocally cause and effect of their form" (CJ § 65, 252). Thus, the first requirement for something to be a natural purpose is that "the possibility of its parts (as concerns both their existence and their form) must depend on their relation to the whole" (ibid.). To sum up, the impossibility for us to conceive natural teleology depends neither on the causation of the whole by the parts (in fact this is nothing new to a mechanistic explanation of nature) nor on the reciprocal relations of parts among each other. Rather, as Förster has brilliantly shown, it is a consequence of the causation of the parts by the whole. This kind of causation is something we know only in one single case: when the idea of the whole precedes and grounds the generation of the parts as their end (Förster 2002, 173). Again we are brought back to a form of anticipation – if not in a temporal sense then in a logical sense of the term – one that reminds us of Toepfer's definition of purposiveness (ZM). It is precisely this anticipation that marks the difference between living organisms and merely mechanical organization: in fact, from the point of view of sheer mechanical laws it is impossible to describe such a form of causation. While the path leading from the parts to the whole or even among parts is perfectly consistent with an explanation based on reciprocal action and therefore locatable within the categories of the intellect, the other path remains inaccessible to the same approach, because it encompasses an internal process of self-generation – i.e. one not guided from the outside. For this reason Kant describes an organism as "both an organized and a self-organizing being, which therefore can be called a natural purpose" (CJ § 65, 253). Unlike inorganic systems, living organisms generate themselves out of their whole, although this is something that eludes our empirical knowledge in a strict sense.

3.2. The dramatic relevance of regulative concepts in Kant's critical philosophy

The reason why a description of living organisms within the methodic framework of natural science is doomed to failure is precisely due to this peculiar anticipation of totality, which is inexplicable from the point of view of efficient causation and yet necessary for understanding organic self-organization. By acknowledging this impossibility Kant marks the frontier, beyond which one should not dare to venture. In fact, in *Critique of Pure Reason* human intellect is supposed to dwell within the limits of the safe island surrounded by a wide and tempestuous ocean (B 294 f.). However, it is precisely in order to keep the business of intellect and hence scientific knowledge both safe and fruitful that limit-concepts (Grenzbegriffe) throw a glance over that very frontier. As Förster shows with reference to Kant's own words, the concept of natural purposiveness (ZM) is necessary for the human faculty of judgment, even if in the form of a subjective principle of reason: "The principle is regulative (non constitutive), but it holds just as necessarily *for our human judgment as it would if it were an objective principle*" (CJ § 76, 288 my italics). This quote is crucial in highlighting the dramatic relevance of "as-if-concepts" in Kantian philosophy: these concepts cannot be in the least reduced to playing a "mere" heuristic or even metaphoric role in science[4]. The limits that mark the frontier imply for Kant always the necessity – albeit perilous – of a glance beyond them.

It is only from a strict neopositivistic point of view that limit-concepts can be reduced to mere "outsiders". We can find an extreme example of this point of view in the ambitious book *Foundations of Biophilosophy* published 1997 by Mahner and Bunge, who intend to provide a philosophical underpinning for biology that is supposed to clear the ground once and for all of confusion and ambiguities. Although the authors do acknowledge a sort of heuristic role of teleology in biology, they restrict its relevance, since "even the heuristic what for-question with regard to an organ x of an

[4] For a metaphorical use of teleological concepts in biology and ecology see a.o. Proctor/Larson 2005.

organism y can be analyzed as a shortcut for a set of nonteleological questions". Further they write:

> "Yet, just as some people, though de facto atheists, cannot give up the notion of religion due to early childhood indoctrination, many biologists and biophilosophers cling to the notion of teleology and try to redefine and re-redefine it" (Mahner & Bunge 1997, 376).

It is evident, however, that the significance of regulative concepts for Kant is much more than the mere psychological need of single individuals for helpful metaphors to describe their research objects. Rather, it is a matter of presuppositions, which might well be external to the field accessible to our discursive intellect, yet are necessary for its very epistemological operations. In fact, Mahner and Bunge turn Kantian as-if conditions upside down by suggesting that biology should pursue its research as if no teleology were at stake. The old scientific criterion of "etsi deus non daretur" – as if there were no God at all – turns thus into an "etsi causa finalis non daretur" – as if there were no final causation at all. On this ground the exclusion of final causation becomes for them a necessary condition for any scientific research at all. This exclusion takes the form of an etsi-non, i.e. a negative condition of possibility.

On the contrary, the Kantian concept of natural teleology is regulative precisely because it is a necessary condition of our knowledge, although it is situated outside of the proper sphere of that very knowledge – intended as empirical knowledge:

> "Hence we must keep to the above principle of teleology – viz. the principle that, in view of the character of human understanding, the only cause that can be assumed [in order to account] for the possibility of organic beings in nature is a cause that acts intentionally, and that the mere mechanism of nature cannot at all suffice to explain these products of nature. But we are not trying to use this principle to decide anything about how such things themselves are possible" (CJ § 78, 298).

Any attempt to describe natural ends in an objective way might very well be for Kant a wild fantasy, since it is impossible to make such a determination within the limits of intellectual knowledge. Yet, any view of nature that reduces all explanation to mere mechanical principles is no less than a wandering among chimeras of natural powers, as Kant clearly states:

> "But reason must not carry this attempt to explain things in mechanical terms to the point of excluding the teleological principle, i.e. to the point of insisting on following mere mechanism even in cases where natural forms are purposive [or specially suitable] for rational investigation into how their causes make them possible and where this purposiveness manifests itself quite undeniably, as the reference to a different kind of causality. For [going to the extreme of explaining everything only mechanically] must make reason fantasize and wander among chimeras of natural powers that are quite inconceivable, just as much as a merely teleological kind of explanation that takes no account whatever of the mechanism of nature made reason rave" (CJ § 78, 296).

Is the Kantian point of view now out-of-date due to his incomplete conception of the faculty of human knowledge? Can we today say that those limits on including final causation within the scientific frontier of mechanical explanation of nature have been overcome? Isn't this peculiar form of causation something that we can very well know with respect to the further development of science? With the exception of some approaches in quantum theory contemporary science insists that all scientific statements refer to causal relations that imply solely efficient causation. This is the case mainly because contemporary science is still grounded on the ontological presuppositions of modern materialistic ontology, according to which matter does not exhibit any kind of internal activity whatsoever and can therefore be fully explained by reference to mere external causal relations.

4. Self-organization theories and the living

Kant's conception of living organisms as self-organizing entities has found a lively resonance within biology and ecology especially after contemporary physics developed self-organization theory and thus made the idea of self-organization amenable to a strict scientific approach.

The theoretical framework of mechanistic-deterministic science can easily include in its explanations reciprocal action and retroaction by referring to non-linear systems characterized by stochastic rather than the classic Newtonian determinism. However, even on this ground – as I claim in this paper – it is still impossible to explain systems and processes that can anticipate the whole as an efficacious end within their internal organization.

Self-organization theories analyze open systems that, as such[5] are distant from the state of equilibrium and therefore follow non-linear processes. Some scholars have considered living beings, which are also open systems since they keep themselves distant from equilibrium, in terms of the general functioning of all open systems, in order to describe life within the methodological patterns of physics (see Prigogine/Stengers 1981, 176ff.; see also Schrödinger 1967, 69ff.).

According to the second law of thermodynamics all isolated systems tend spontaneously to fall back onto the closest equilibrium state, i.e. they tend to a rest status, to the maximal possible entropy relative to that system. However, if its energy increases the system runs out of balance and reacts with a process that brings it back to rest. If the system is quite close to equilibrium this process follows a linear path and quickly dampens down interferences. However, when the system is very far from equilibrium and interferences have reached the threshold proper to that system, a somewhat more complex process occurs, which is characterized by extremely high fluctuations and strong unpredictability (Koutroufinis 1996, 24ff.). In this state any feedback operates as an amplifier that extends fluctuations to the whole system. In this respect open systems do indeed resemble living beings, since both incessantly generate disorder, which they must transfer to their environment in order to keep internal order.

So-called self-organized systems are driven by their tendency towards equilibrium, from which they have to be constantly kept away. It looks pretty much as if they were all the time struggling to return to the quiescent state. In this "struggle" against outer interferences, which keep open systems from equilibrium (because, being open, they are exposed to them), so-called self-organized systems generate entropy by their inner processes (such as chemical reactions) and tend to go back to maximal entropy. This

[5] Generally speaking scholars make a distinction between isolated systems, in which no flow of energy or matter takes place, and closed or open systems, in which energy flow and in the latter case even matter flow take place. I use here the term "openness" (being open) in a technical sense to include both closed and open systems.

is why they are termed by Prigogine "dissipative structures". In order to generate as much entropy as possible per time unit they have thus to organize themselves (Koutroufinis 1996, 20ff.). By doing so they originate complex structures[6].

The concept of self-organization refers to the behavior of a system that, once it has reached a critical threshold, starts organizing by itself (see Schneider/Kay 1995)[7]. The term "by itself" means that there is no need for any external agency, be it a force or a designer or even a program, in order to explain this process (see Fox Keller 2005). Accordingly, self-organized systems can spontaneously achieve a dynamic stability out of a disordered status without falling back into thermodynamic equilibrium.

Is this the case for a kind of internal teleology which biology might include constitutively in its explanation of living beings? Certainly we are faced here with a peculiar form of reciprocal action that, while neither questioning nor suspending natural laws, still pushes them to their explanatory limits. Therefore, self-organization theories break away from classic determinism and offer a far better ground for the analysis of open systems complexity than traditional physics. No wonder then that these theories meet with great approval in biology and ecology.

However, at a second glance, it becomes clear that the only teleological act in so called self-organized systems in a laboratory is performed by the researchers themselves as they supply parameters as well as initial and constraint conditions and adjust them to regulate the course of the process. By targeted interventions, which keep the systems constantly away from equilibrium, researchers can exert some control over the process and its ef-

---

[6] Defining complexity is a very difficult task because the term is used in different ways by different disciplines. From the point of view of physics complexity is linked by inverse relation to predictability (the more complex the less predictable). According to Goodwin and Solé complexity is a dynamically stable state between complete order and complete disorder, i.e. it wavers on the edge of chaos (Goodwin/Solé 2000, 33). At this status a system can achieve the highest possible differentiation without collapsing into sheer chaos. In this respect self-organized systems are complex.

[7] By quoting Bak's definition of "self-organized criticality" Evelyn Fox Keller presents the well-known example of a sand pile to describe a model of an open and dissipative system. A sand pile retains its conical shape as more sand is added by means of "self-organization", i.e. by setting off small avalanches at the critical point (Fox Keller 2005).

fects of feedback looping. At the same time, researchers fix the boundaries of the system for methodological reasons in such a way that they themselves are excluded from it. Therefore, since the only truly teleological element of the system – i.e. the acting "subject" – is ejected from the explanation of it, it is possible to exclude teleology from that very system altogether. As Fox Keller maintains, the "self" is thereby cleared out of self-organization[8]. Nonetheless, the main difference between a living system and an inorganic "self-organized" system is represented precisely by the active intervention of the researcher. In fact, the only "spontaneous" activity of inorganic self-organized systems is its strenuous attempt to return to quiescence. Such systems by no means tend to spontaneously generate high complexity, which is possible only in a status very distant from equilibrium. The fact that complexity nonetheless arises can be considered as a byproduct of this constant effort to reach quiescence. This is why such systems can be explained fully within the framework and the jargon of natural science based on non-linear reciprocal action without any reference to teleological expressions at all.

Living beings, on the contrary, defeat such an explanation. In fact that very "self" that, in the case of inorganic self-organization is represented by researchers who constantly interfere with the system in order to have it generating complexity, seems to be at work in living beings as well, yet in the absence of any researcher! Moreover living beings seem actively to look for "interferences" that obstruct a falling back onto quiescence and that keep them in a steady dynamic process. On the one hand, living beings constantly generate high complexity by seeking stimuli that bring them out of balance. On the other hand, in spite of the ever-increasing complexity, they manage to restore internal order. In other words, living beings attain a high grade of internal coherence that combines the achieved complexity with relatively dynamic stability. This is possible mainly because living beings regulate their being open or closed towards the environment on their own: thus, they may look for stimuli and thereby keep themselves as far as possible from the

[8] "To bring the engineer into the system, to put Levin in his office with his agency and intentionality intact, would be to confound the entire tradition that takes human agency or intentionality as a priori unnatural, and accordingly pits natural against artificial design" (Fox Keller 2005). This reminds also of the title of Koutroufinis' book: "Self-organization without self" (Koutroufinis 1996).

status of maximal entropy and at the same time uphold their inner coherence and avoid collapsing (Koutroufinis 1996, 108). All that is performed by researchers or (in the case of a self-organized system independent from laboratory conditions, like a tornado) external conditions by means of induced amplifying feedback-loops, is carried out, in the case of a living being by itself.

The adoption of the term "self" in the sense stated above should not be confused with any vitalistic concept. According to Vitalism and Neovitalism a new principle like a soul or an entelecheia is added to the otherwise merely materialistic-mechanical explanations of physical laws. This principle is supposed to lead natural processes in a non-materialistic way and does not directly obey the general physical laws of nature. The term "self" introduced in this chapter is not at all a supernatural principle. Rather, it stands for the self-regulating center that is necessary to explain the unique interconnection between growing complexity and generation of inner order that we find in living beings.

It is indeed extraordinary how living beings manage to preserve their inner coherence in spite of the ever-increasing complexity they constantly generate by means of energy input. From the point of view of physics it is puzzling how they manage not to die, i.e. not to fall back into thermodynamic equilibrium just as the so-called self-organized systems under laboratory conditions would do if left on their own. As Koutroufinis shows, in order to explain such an incredibly high generation of coherence we have to allow for a kind of anticipation that regulates the relations between parts as well as between parts and wholes (1996, 119ff).

Living processes can only occur over an extended time span. In doing so, they regulate their openness towards the surrounding environment from within and keep their boundaries – membrane-like or not – selectively permeable (Koutroufinis 1996, 108). This means that, in order for inner coherence to be carried out, a form of anticipation of future states is strictly needed. Such an anticipation encompasses not only the internal structure of a single living organism, but has also to stretch out to the environment relevant to that very organism – that is the environment with which it entertains a permanent interaction.

By the consideration of the necessity of anticipation for living beings we are brought back to Förster's interpretation of Kant: the logical anticipation of the respective whole, which is never given once and for all due to its being constantly in progress, is a necessary condition for preserving inner coherence. Of course the whole cannot precede the parts and act on them in the form of efficient causation, since it is generated by those very parts. Yet it must precede them – as Förster maintains – as their "end" in order for them to engender it at all.

So far it is very difficult to say whether one can already introduce the concept of intentionality, i.e. whether living organisms intentionally preserve their very conditions of surviving a highly probable collapse into death. As Kant has shown, an answer to this question can only take place outside the realm of natural sciences and their methodic approach. However, following here Evelyn Fox Keller, if we are to drop "intentionality" we might very well adopt the term "agency" intended as something that "we clearly share with many if not with all other organisms" (Fox Keller 2005). For this reason we need an additional distinction to those offered by Toepfer in his analysis: the distinction between intentionality (ZS) and agency (ZT). While the first term refers to that particular form of intentional orientation towards an end that is the ground for practical action (Handlung) and presupposes free will as well as consciousness (see Krebs 1997, 352ff.), the term agency (ZT) as a necessary condition for the preservation of living organisms implies logical and temporal anticipation and therefore also a certain kind of differentiation between possibility and reality, without requiring the high performance of a human-like mind.

In the latter case we are faced with a peculiar form of causality, which indeed – as Kant says – has no analogy to any causality that we might know on the ground of our empirical possibilities of explanation. Such a form of causality corresponds to what Fox Keller calls agency and cannot be explained according to reciprocal action, as it is the case instead for purposiveness (ZM).

If we decide to stay within the Kantian framework we have to admit that the only way out of the impasse of purposiveness (ZM) is to consider it as a limit-concept in our description of nature. Therefore, empirical natural sciences must at least allow for a kind of teleology of nature within their methodological field as a necessary epistemological as-if condition, if they are concerned with living beings and their interaction.

The question about a constitutive use of teleological concepts cannot be asked within the framework of critical philosophy, since Kant's discursive intellect cannot properly encompass teleology (Förster 2002, 176ff.). Addressing this topic is, however, a task for metaphysics or ontology[9].

5. Whitehead's ontological underpinning for teleology

5.1. Supersensible is not the same as supernatural

According to Kant a reconciliation between efficient causation and final causation in nature is impossible, since the two kinds of explanation exclude each other:

> "If we are to have a principle that makes it possible to reconcile the mechanical and the teleological principles by which we judge nature, then we must posit this further principle in something that lies beyond both (and hence also beyond any possible empirical presentation of nature), but that nonetheless contains the basis of nature, namely we must posit it in the supersensible, to which we must refer both kinds of explanation" (CJ § 78, 297).

Whitehead would plainly agree with the assertion that final causation lies beyond our empirical representation of nature, as far as the latter is understood in terms of mere sense perception and its extension through laboratory experiments. This path, which natural sciences follow, is rooted in the ontological presupposition that Whitehead calls "scientific materialism"

[9] According to Förster, the possibility of an intuitive intellect, as far as it is intended as synthetic-universal intellect and not as the world-causation principle (Weltursache), cannot be completely excluded from a Kantian framework. He shows that allowing for intellectual intuition does not lead to a contradiction with Kantian thought, since intellectual intuition does not necessarily imply creativity in the sense of a creatio ex nihilo ("produktive Einheit von Möglichkeit (Denken) und Wirklichkeit (Sein)" – the productive unity of possibility (thinking) and actuality (being)). Rather it can refer to a "sense intuition of things as such" (sinnliche Anschauung von Dingen an sich) (Förster 2002, 179). Accordingly, we can allow for an access to a teleology in nature still within the limits (the boundaries) of pure reason. Förster argues that it was Goethe who followed this very path.

(1967a). Empiricism based on mere sense perception has ended up constituting the core of any scientifically sound knowledge.

However, according to a Process framework, sense perception, far from being the first immediate mode of relation to an objective world, is the outcome of a very complex process of symbolic reference and embodies only a fraction of what experience more broadly means[10].

For Whitehead as well as for Kant, final causation lingers in the fields of the supersensible, since it cannot be directly perceived by the senses and their technological prostheses. According to Whitehead, however, supersensible is not identical with supernatural: therefore, considering final causation beyond the boundaries of sense perception does not lead to the introduction of supernatural concepts. Unlike Kant, Whitehead does risk entering that nebulous ocean, which lies beyond the safe boundaries of knowledge secured by the link between sense perception and intellectual concepts. For that ocean is the very place in which we originally dwell, whereas sense perception and its intellectual organization by the categories of modern sciences represent only a tiny island, on which we can settle down. By pushing Kant's metaphor a little further we can say that the boundaries around that island rely on specific ontological presuppositions, which Whitehead terms the materialistic ontology of substance (and permanence)[11].

5.2. Ontological background

By definition Whitehead's philosophy of organism aims at reconciling Kant's two forms of causality such that neither can exist without the other. Accordingly, any theory addressing questions of ontology and natural philosophy cannot avoid a co-implication of both forms of causality without falling into a logical contradiction:

> "[T]he doctrine of the philosophy of organism is that, however far the sphere of efficient causation be pushed in the determination of components of a concrescence – its data, its emotions, its appreciations, its purposes, its phases of subjective aim

---

[10] For a detailed analysis of Whitehead's concept of experience see Whitehead 1985.

[11] For a critique of this paradigm within a process framework see Hampe 1990.

> – beyond the determination of these components there always remains the final reaction of the self-creative unity of the universe. This final reaction completes the self-creative act by putting the decisive stamp of creative emphasis upon the determinations of efficient cause" (Whitehead 1978, 47).

By talking about "the self-creative unity of the universe" Whitehead does not mean a general point of view over the whole universe as a given. Rather, he refers to a process of growing together, by which past efficacious influences are prehended and brought into the unity of a novel shape. This process is termed by Whitehead concrescence and means at the same time growing together (from the Latin cum-cresci) and becoming concrete. Since efficient causation alone cannot explain the existence of an actual world, the introduction of final causation is an ontological and logical necessity, as I will show below.

In his early philosophical works Whitehead engaged in a careful analysis of the concept of time according to classical physics and its contemporary critique accomplished by the first steps of quantum as well as relativity theory. For the sake of brevity it is impossible to follow up this development of Whitehead's philosophy. However, it is from this early analysis that he came to the conclusion that nature can be thought of only in terms of passage and therefore permanence can only be considered as a result of becoming[12]. The fact that there is something rather than nothing is not at all apparent on the surface. Rather, actuality and permanence are the outcome of a relentlessly creative activity. Thinking of this activity in terms of a permanent flowing would lead to a monistic metaphysics, according to which a single all-encompassing actual entity (reminiscent of Spinoza's concept of substance as the only ens realissimum) would bring about forms within itself. Against this position Whitehead claims for a plurality of occasions of experience as entes realissima (indeed actual entities in the most proper sense of the term), which do not change and do not

---

[12] By broadening the traditional concept of extension from space to time, i.e. abandoning the (modern) conception of time as a series of separated instants, whose connection is hard to explain, we have to allow for extended overlapping events. This overlapping connectedness constitutes the core of the idea of a 'passage of nature', from which the physical conception of time is then derived. Whitehead describes this very connection as a causal one (cf. 1920, Emmet 1984).

move, but occur and perish. The permanent existence of complex structures, which Whitehead terms societies, is the outcome of processes of actualization and repetition of patterns, which are brought about by actual entities. Once we give up permanence as the most originary mode of being, we have to follow Whitehead in conceiving reality in terms of actualization. Rather than recalling the idea of a creatio ex nihilo this process is always a conditioned self-shaping activity, in which possibility turns into actuality. At this point one main difference from Kant's philosophy becomes very clear: Whitehead attributes reality strictly speaking to becoming (as actuality (Wirklichkeit)) and only in a derivative sense also to being, which can be strictly conceived only as having already 'been'.

Possibility can be conceived within a Whiteheadian framework in terms of mere possibilities or as real potentialities for actualization. Accordingly, timeless and abstract possibilities are subject only to logical restrictions and hence represent mere possibilities for actualization. These are termed by Whitehead eternal objects, since they are existing but not actual, therefore neither efficacious nor temporal. In order to become actual they need an efficacious act that brings them about. Whitehead does not ascribe this agency to a single entity (like Kant's Weltursache) but to a plurality of occasions of experience. Consequently, Whitehead's actual entities are to be considered in a strict sense as agencies that – to use a metaphor – weave the reign of possibilities into the texture of reality (indeed it is patterns which are actualized).

To say that an actual entity is an agency does not imply that it is severed from any ties and therefore absolutely free. While actual entities have an external relation towards the present (contemporary entities) as well as towards the future and are therefore free from this point of view, their relation to past entities is internal and hence causally determined. This is what Whitehead means when he refers to real potentialities, which are also open to further realization; yet, differently from eternal objects, they are neither timeless nor inefficacious (1967b, 179; see also Cloots 2000). Past entities, which have already actualized them, are an inescapable condition for future actualizations. Although past entities have completed their process of becoming and have hence "perished", due to their causal efficacy they are still active upon future actualizations that objectify them[13]: "The 'objectifications' of the actual entities in the actual world, relative to a definite actual entity, constitute the efficient causes out of which that actual entity arises" (Whitehead 1978, 87). Past entities as the 'given' past world constitute the very presupposition for the occurring of new entities. Therefore, they exert a certain kind of power in each novel actualization: "In Locke's phraseology the objectified actual entity is then exerting 'power'" (ibid. 58). As real potentialities they restrict the self-creative activity of each new becoming entity. In several writings Whitehead argues against the idea of an absolute freedom:

> "[T]here is no such fact as absolute freedom; every actual entity possesses only such freedom as is inherent in the primary phase 'given' by its standpoint of relativity to its actual universe" (ibid. 133).

Therefore, each actual entity is a conditioned agency that turns possibility into actuality.

Causal efficacy is a necessary but not sufficient condition for conceiving the passage of nature without logical contradiction. If we are to follow Whitehead in considering a pluralistic universe, in which all past entities exert causal efficacy upon the future, we are faced with the difficulty of explaining how it is possible that novel forms arise. Let us consider the influences of past entities in terms of vectors, as Whitehead himself allows for[14]; in this case we are led to envision a incredibly interwoven web of

---

[13] Whether past entities are to be considered merely passive or still active is a central and live issue among Whitehead's scholars. As Nobo clearly shows, the term 'actual' in Whitehead's work does not exclusively refer to new becoming entities, but also to the perished ones (see Nobo 1974). Accordingly, past entities have very well perished, since they do not present subjective activity anymore. However, they are not merely passive, as Frankenberry also maintains: "Satisfaction spells the death of the process of unification but not the end of the creative energy involved" (Frankenberry 1983). Past entities are hence efficacious forces that throw themselves into the novelty of the future: "the throbbing emotions of the past hurling itself into a new transcendent fact" (AI 227) (See for this discussion also Muraca 2005).

[14] According to Frankenberry "physical prehensions are 'vectorial' designating transmission of energy with direction as well as magnitude". They are "the physical transfer of energy" (Frankenberry 1983).

threads, which thwart and neutralize each other. In the absence of a converging force they would lead to "cross-current incompatibilities" (ibid. 247). Virtually endless causal vectors cannot automatically bring about a novel shape (see Muraca 2005), as Whitehead himself points out:

> "[E]ach novel actuality is a new partner adding a new condition. (...). Each condition is exclusive, intolerant of diversities; except so far as it finds itself in a web of conditions which convert its exclusion into contrasts" (PR 223).

This 'web of conditions' cannot be the mere outcome of fortuitous connections among vectors. In order for causal efficacy to contribute to the generation of novel shapes we need a point of convergence for all relevant vectors that attracts the causal influences and orders them by selection and canalization. The process, in which a coherent unity arises out of a variety of several conditions that tend to thwart each other, necessarily needs a kind of orientation from its not yet actualized completion. This is precisely that anticipation of (provisional) totality, which we already introduced earlier. Without this anticipation, which filters past influences by means of abstraction and lets them grow by means of canalization from their initial incompatibilities into rich – and beautiful – contrasts, no coherent shape can take place. Such anticipation cannot be conceived within a process of concrescence as temporal in a strict sense of the term, since time is generated by that very process itself.

The anticipated completion (totality) works as the point of concentration and selection of past influences. This very point, in which the so-called subjective aim of each actual entity crystallizes itself, is both the aim and the result of the concrescence process and leads that very process of becoming towards itself by attracting it as an erotic 'lure' (see Muraca 2005, 241)[15]. The subjective aim is the point of attraction of all vectors and therefore it is the "lure for feeling" (Whitehead 1978, 85). The subjective aim is by no means something already 'given'; rather, it can be construed as a kind of force of desire or as a longing for realization that is still vague

[15] The very first impulse for a concrescence process is given by God's so-called initial aim. Actual entities can then consider the initial aim differently (by conforming to it or even rejecting it) and turn into their subective aim.

and undetermined in its starting phase and grows then towards the aim of satisfaction.

How past influences are felt and combined, with what intensity they are prehended, which ones are more relevant than others, in other words, how a new shape occurs, is determined by the free acting of each actual entity in its subjective form. The degree of novelty that can be achieved depends on the level of internal complexity of that very entity.

Kant was right in acknowledging that only a creative intellectual intuition freed from sensibility can turn possibility into actuality without having to rely on the practical Sollen. According to Kant the Sollen is the only way of connecting possibility to actuality for finite beings that are not endowed with intellectual intuition. However, what Kant did not sufficiently consider in his work is the other way of bridging the gap between possibility and actuality, one wholly appropriate to finitude: that of desire and lure. Through them future possibilities for realization can be anticipated in the present and desired with a strong emotional investment. Unlike Kant, Whitehead roots his metaphysics on an aesthetic or even (in Plato's sense of the term) an erotic principle in order to elucidate the actualization of possibility.

5.3. Teleology as a necessary postulate

Final causation is an ontologically necessary assumption and in its concrete figures it is inaccessible to our perception. It is precisely because the principle of final causation manifests itself within the process of concrescence of each actual entity and not, as is the case in efficient causation, in the transition between entities, that it eludes observation. Since final causation cannot become an object of perception, it can only be assumed as a postulate. Final causation takes place in the private phase of a becoming process, that is a phase which is locked to causal influences from contemporary entities. As we know from *Process and Reality*, contemporaneity is defined in terms of causal independency: two entities, which are both in their becoming, cannot influence each other, because technically they are not yet and therefore they cannot be efficacious. Once the becoming process is accomplished, what is discernible is no longer the private self-constituting process, but only the entity as its very result, i.e. its final shape and the patterns actualized in it.

In other words, by looking back into the past of already actualized entities, we recognize the realized patterns, but cannot detect any caesura in them, which might somehow indicate the single processes of actualization. Once the texture of the world has been actualized by all single concrescence processes, it looks like a continuous structure, which can be the object of observation. This very structure can be thus divided arbitrarily by any observer, since the actualized past (which is 'being' qua already occurred) is perceivable as a continuum, in which the division process never leads to a zero point or a smallest indivisible element[16]. Actual entities are by no means the smallest constituents of the universe, even if Whitehead sometimes describes them as microcosmic organisms. At the end of their actualization process they are no longer locatable as teleologically acting subjects, since they are now "objectively immortal" and therefore acting only as efficient causes. The concretum, i.e. the world actualized by processes of concrescence, is divisible independently from the subjective unity of the concrescences that generated it[17]: "In dividing the region we are ignoring the subjective unity which is inconsistent with such division" (1978, 284). In contrast, the genetic process of an actual entity is not divisible,

[16] "The actual world is atomic; but in some sense it is indefinitely divisible" (PR 286). With the term ‚actual world' Whitehead intends both the becoming world within the contemporaneity of unison of a duration (and this is atomic, indivisible and unperceivable, since 'private') and the already actualized world. This twofold definition sounds somehow confusing and can be understood only on the ground of the distinction between actuality and potentiality: from this point of view both the becoming entities and the past world are actual (cfr. Nobo 1974).

[17] Whitehead's extensive continuum refers to the potentiality for actualization, since the possibility of extension is a necessary assumption for the connectivity of all overlapping actual entities. Given the extensive continuum as a potentiality, this very connectivity is then actualized only by the internal relations among actual entities. Thus Whitehead can say that each concrescence presupposes its region, whereas the region is the mere topological potentiality of actualization and is therefore indifferent as to its divisibility. The atomization of the region is a consequence of a process of actualization. Once the process is accomplished the region can be divided regardless of that very atomization.

since it does not take place within time and occurs all at once: "Each phase in the genetic process presupposes the entire quantum" (ibid. 283).

We can hence conclude that actual entities can be by no means for Whitehead the object of sense perception or direct observation. All that we can perceive is the nexūs, which consist of repeated patterns and actualized relations between entities that have completed their process of becoming.

We have shown within a Kantian framework how teleological concepts in the form of "as if" conditions are an epistemological necessity for any scientific approach that attempts to analyze life. With Whitehead's help we can then venture a step beyond Kantian limit-concepts into ontology, by showing that teleology in terms of inner agency (ZT) is a necessary ontological assumption, in order to be able to think actuality at all. At the same time, we have also illustrated that the specific form in which that very agency (ZT) unfolds itself eludes our possibility of experiencing it directly.

Nonetheless, that very agency (ZT) that at the ontological level remains inaccessible to experience, reaches a point of relevant manifestation with living beings and therefore at this very level it can at least be intuitively grasped.

5.4. Societies: different degrees of relevance for agency (ZT)

All actual entities swing between generation of novelty and repetition of past patterns. As Whitehead in *The Function of Reason* writes, there is on the one hand a general tendency to decay, which at the same time works as a factor of stabilization for realized patterns and hence provides order. This tendency is 'methodical', it is inclined to repeat given configurations, and supports thereby the permanence of common structures. The ordering tendency is rendered possible by abstraction. Abstraction is intended at first as an ontological process of exclusion and neglect of all past influences, which are considered non-relevant or even destabilizing. It is by means of abstraction that the high complexity of efficacious factors can be reduced to a necessary minimum of relevant and non-dangerous influences. However, if this tendency reaches too strong a degree, no novelty will take place. When a complex structure like a society does not open itself to novelty and 'remains' the same for a long time with respect to its patterns, a slow process of decay sets in. In a non-permanent world, in which becoming is the source of being, there is no everlasting stability. This decay corresponds to what in physical terms is called an entropic process.

The other tendency works of course in the very opposite direction: it is anarchic, it breaks patterns, and it generates novelty. It opens to more complexity and at the same time to a risky balance among all inflowing factors. Yet, it is the condition for the emerging of complex structures like living organisms. Being open to more possibilities of realization (in this case both real and mere potentialities) can elicit complexity while also being exposed to the danger of it.

We can distinguish actual entities according to the degree of novelty that they can generate. This depends both on their degree of abstraction, i.e. exclusion of past influences as well as on their capacity of canalizing incompatibilities into contrasts. However, the generation of novelty does not only depend on each single concrescence process at the fundamental-ontological level. The entities we normally deal with and the objects we perceive are obviously not actual entities; rather, they are the outcome of relations among actual entities and these relations can take a more or less hierarchical or loose structure. At this level the organizational form of a group of occasions is the crucial factor for the very relevance of whatever novelty can be generated at a fundamental-ontological level. In other words, whether novelty becomes a relevant factor at the next organizational stage, in which occasions of experience group together, is determined by the form that the organization takes. In some cases any attempt at novelty is averaged out, whereas in others it reaches the core of the whole grouping. The most general way of grouping of occasions is termed by Whitehead a nexus, which is a group considered merely with respect to the basic property of mutual immanence while otherwise lacking in common relevance (1967b, 201). A nexus that enjoys some kind of social order is termed a society and is instead characterized by a common element of form, which is illustrated and reproduced by each actual entity that constitutes that very society (ibid. 203). All the members grouping in a society actualize the same or alike patterns precisely because their connection is tighter, since "by reason of their common character, they impose on other members of the society the conditions which lead to that likeness" (ibid. 204; 1978, 89). In the case of a society the order is guaranteed by the commonality of patterns repeated with certain continuity. Non-living aggregates like stones, which are termed by Whitehead 'corpuscular societies' (1978, 35), are the outcome of the average reciprocal influences of their members: "For lifeless matter these functionings thwart each other, and average out so as to produce a negligible total effect" (1967b, 207). In this case any novelty that might have been generated at the fundamental-ontological level of the entities constituting a society will disappear under the veil of the average behavior of all other members of that society; thus novelty will be driven into sheer irrelevance. The law of averages protects the society from the risk that novelty brings about, i.e. the risk of a chaotic falling apart.

In the case of living beings we are confronted with a completely different organizational form and behavior: the teleological drive constitutes an essential character of their very existence. This means both a highly relevant realization of complex novelty and the capacity to coordinate and stabilize it. In living organisms the self-shaping activity breaks through at the level of the whole society and thereby it gains visibility.

According to Whitehead we can speak about life only at the stage of complex grouping of occasions, since life is not a peculiar characteristic of a single actual occasion. Rather, it refers to a specific mode of coordination among the several spontaneous and creative activities of all members of a society, which no longer thwart each other like in lifeless matter, but are directed towards a common aim: "The essence of life is the teleological introduction of novelty, with some conformation of objectives" (ibid.).

5.5. Life as a catalytic agency

The easiest way of achieving permanence is the path of abstraction, i.e. the exclusion of all influences that might somehow hinder the average reproduction of patterns:

> "[E]liciting a massive average objectification of a nexus, while eliminating the detailed diversities of the various members of the nexus in question. This method, in fact, employs the device of blocking out unwelcome detail" (Whitehead 1978, 101).

This is precisely the strategy that non-living societies follow. It is a safe path that does not, however, allow for much complexity, which would rather jeopardize the stability of the whole society.

While more complex societies can allow for more influences and possibilities to actualization to flow in, they tend to be less stable and more exposed: "a high grade of complexity will in general be deficient in survival value" (ibid. 101). In other words, the chance of a development towards increasing complexity risks dissolution. This is also true for species, as Whitehead points out; when a species tries to leap to a higher level "it can stabilize itself, and relapse so as to live; or it can shake itself free, and enter upon the adventure of living better" (1929, 19). However, the attempt might end up either way: "If the choice be happy, evolution has taken an upward trend: if unhappy, the oblivion of time covers the vestiges of a vanishing race" (ibid.).

Life is the outcome of a successful strategy of combining high complexity with survival and enduring. A living society manages to gather novel elements from its environment and to reconcile them with the experiences of its members. In this way a society can elicit novelty in order to match the novelty of its environment, instead of defending itself from it by means of abstraction (1978, 102). In other words, living beings canalize the wide range of influences into a new, original and relatively stable shape:

> "Apart from canalization, depth of originality would spell disaster for the animal body. With it, personal mentality can be evolved, so as to combine its individual originality with the safety of the material organism on which it depends" (ibid. 107).

Due to their wider openness living beings rely much more on a constant exchange with their environment than non-living ones in order to reestablish at each step a provisional stability, which is constantly thrown out of kilter by novel influences. Living beings clearly dwell at some distance from the law of averages that supplies other societies with endurance. They are constantly at risk of slipping out of their fragile balance between novelty and surviving:

"[T]he structure is breaking down and being repaired. The food is that supply of highly complex societies from the outside which, under the influence of life, will enter into necessary associations to repair the waste" (ibid. 106).

Rather than being a characteristic or an attribute of enduring entities, according to Whitehead life is a kind of activity, a sort of quasi-agency that cannot be described objectively, but becomes manifest through its efficacious behavior. Life coordinates, canalizes, and catalyzes: "life acts as though it were a catalytic agent" (ibid. 106).

A society can be termed living when so called entirely living nexūs are *dominant* in it (in Whitehead's terms: *regnant* ibid. 103). Whitehead describes an entirely living nexus in terms of a nexus whose members attain a very high degree of novelty, while the nexus itself is said to be non-social (ibid. 107). In comparison to another nexus an entirely living nexus seems to be a quite isolated knot that can generate a high grade of novelty precisely because its dependency on a serially ordered society is not strongly relevant. Somehow the continuity of repetition of patterns is partially suspended; new variables are tested as if it were something like an enclosed laboratory. However, this does not lead to the conclusion that the members of an entirely living nexus are completely independent from any causal influence. Rather, "each member of the nexus derives the necessities of its being from its prehensions of its complex social environment" (ibid.). The point is a relative independence of a nexus from the privileged context of extremely relevant influences that constitute a typical society: "By itself the nexus lacks the genetic power which belongs to ,societies'" (ibid.).

On the one hand life needs such a laboratory for novelty, but it is much more than a laboratory: otherwise the novelty forged in the lab would simply lead the whole organism, in which it arises, to collapse.

Indeed entirely living nexūs would never have a chance of surviving if they were not surrounded by subordinate – most of the time inorganic – other nexūs that protect them and limit their high self-shaping freedom for the sake of stability:

"A complex inorganic system is built up for the protection of the 'entirely living nexus', and the originative actions of the living elements are protective of the whole system. On the other hand, the reactions of the whole system provide the intimate environment required by the 'entirely living nexus'. We do not know of any

> living society devoid of its subservient apparatus of inorganic societies" (ibid. 103).

Therefore, living beings are constituted by living as well as non-living nexūs, which constantly reestablish the balance between order and disorder.

From this point of view an entirely living nexus cannot be utterly non-social with respect to the relevance of the society that surrounds it: "[T]hough life in its essence is the gain of intensity through freedom, yet it can also submit to canalization and so gain the massiveness of order" (ibid. 107). Entirely living nexūs act as catalytic factors for freedom and novelty: although they cannot generate full novelty for the whole society on their own, they still can intensify its tendency to the generation of novelty.

We might then say that due to the missing social connection of a particular nexus life as a catalytic agent can somehow lure novelty; life dwells in the gaps of a tightly interwoven web of relations, as Whitehead himself writes:

> "In a nexus of living occasions, there is a certain social deficiency. Life *lurks in the interstices of each living cell*, and in the interstices of the brain" (ibid. 105-106).

Given the unstable balance and the high complexity of living organisms, anticipation plays a major role in guaranteeing their inner coherence:

> "[I]n so far as the relevant environment is dominated by any uniform type of coordination, any occasion will experience its past as ‚anticipating' its prolongation of that type of order into the future beyond that past" (1967b, 196).

Since future occasions do not exert any causal influence on past ones, in absence of anticipation they would be completely neglected by the self-shaping process of becoming entities. Whenever relevant novelty is generated the risk of dissolving the established order would be extremely high. Therefore, the coherence that is evident and necessary in all living beings can be explained only on the ground of a high performance in terms of anticipation, by which an immanentization of the future into the present takes place:

> "For, apart from contemporaries, one occasion will be in the future of the other. Thus the earlier will be immanent in the later according to the mode of efficient causality, and the later in the earlier according to the mode of anticipation" (ibid. 197).

Without such anticipation it would be impossible to explain the strong cohesion of a society that achieves a high complexity by the relevant activity of its entirely living nexūs.

6. Conclusion

In the first part of this paper I presented with reference to Toepfer's classification different concepts of teleology, which are relevant for the life sciences; I then showed how only agency (ZT) and intentionality (ZS) can be considered as truly teleological concepts. Agency (ZT) does not necessarily imply intentionality, but only the faculty of anticipation. In fact all the other definitions that scholars include in the realm of teleology (internal and external purposiveness (ZM)) can be explained in terms of mere efficient causality and are therefore not alien concepts in the methodological approach of the life science. This is even truer after the latest developments of contemporary physics with respect to highly complex theories of (efficient) causality, which reach far beyond the sheer causal linearity of classic theories. In spite of their being extremely useful for biology and ecology – something we should not forget – these theories still cannot internalize teleological concepts within the methods of science, since teleology in a strict sense lies beyond the limits of accessibility of empirical scientific approaches.

By considering Kant's *Critique of Judgment* I then showed that natural teleology cannot simply be reduced to a mere explanation in terms of efficient causality, even if the latter is intended as a complex form of reciprocal action. Hence, teleology is and remains a necessary limit-concept for science as far as its approach is rooted in what Whitehead has termed scientific materialism and its presupposition of mere external efficient-causal relations among phenomena. To say that teleology is a necessary limit-concept is not to underestimate its crucial role for science or to exclude it from the field of consideration in which science dwells. Rather, by reference to Kant's exposition about limit-concepts I have shown how dramatically relevant they are for any scientific attempt at describing and explaining nature, especially in the case of living beings. Therefore, I claim that a heuristic-regulative (in Kant's radical sense of the term) employment of teleological language represents a minimum requirement for science.

Furthermore, as I have illustrated, an ontology that can offer an argumentative underpinning for teleology without having to make a leap into supernatural categories can venture a step beyond the Kantian criticism. Drawing on Whitehead I showed that teleology must be a necessary ontological presupposition for thinking actuality at all. With respect to the understanding of living beings this presupposition is especially significant. Whitehead's philosophy of organism offers a plausible ontological ground for self-organization theories that dare to venture beyond their limits and presuppositions.

REFERENCES

Cloots, A. (2000). "The Metaphysical Significance of Whitehead's Creativity". In: *Process Studies*, 30, (1, 2000), pp. 36-54.*

Emmet, D. (1981). "Whitehead's View of Causal Efficacy". In: Holz, H.; Wolf-Gazo, E. (eds.): *Whitehead und der Prozessbegriff: Beiträge zur Philosophie. Alfred North Whiteheads auf dem ersten Internationalen Whitehead-Symposion 1981* [= "Whitehead and the Idea of Process: Proceedings of the First International Whitehead-Symposium 1981"]. Freiburg: Alber (1984), pp. 161-178.

Fox Keller, E. (2005). "Ecosystems, Organisms and Machines". In: *BioScience*, 55, pp. 1069-1074.**

Förster, E. (2002). „Die Bedeutung von §§ 76, 77 der Kritik der Urteilskraft für die Entwicklung der nachkantischen Philosophie". In: *Zeitschrift für philosophische Forschung*, 56, (2, 2002), pp. 169-190.

Frankenberry, N. (1983). "The Power of the Past". In: *Process Studies*, 13, (2, 1983), pp. 132-142.*

Hampe, M. (1990). *Die Wahrnehmungen der Organismen. Über die Voraussetzungen einer naturalistischen Theorie der Erfahrung in der Metaphysik Whiteheads*. Göttingen: Vandenhoeck und Ruprecht.

Kant, I. (1998). *Critique of Pure Reason* (Translated and edited by Paul Guyer and Allen W. Wood). Cambridge, New York: Cambridge University Press.

—— (1987). (CJ). *Critique of Judgment* (Translated by Werner S. Pluhar). Indianapolis: Hackett Publishing Company.

Koutroufinis, S. (1996). *Selbstorganisation ohne Selbst. Irrtümer gegenwärtiger evolutionärer Systemtheorien.* Berlin: Pharus.

Krebs, A. (1997). *Naturethik – Grundtexte der gegenwärtigen tier- und ökoethischen Diskussion.* Frankfurt am Main: Suhrkamp.

Mahner, M.; Bunge, M. (1997). *Foundations of Biophilosophy.* Hamburg: Springer.

McLaughlin, P. (2005). „Funktion". In: Krohs, U.; Toepfer, G. (ed.) (2005): *Philosophie der Biologie.* Frankfurt am Main: Suhrkamp, pp. 19-35.

Muraca, B. (2005). „Wie kann sich etwas, was noch nicht ist, sich aus seiner Zukunft heraus frei gestalten? Identitätsbildung zwischen Kausal- und Finalursache ausgehend von Whiteheads Kreativitätsbegriff". In: Abel, G. (ed.): *Kreativität. XX. Deutscher Kongress für Philosophie. Sektionsbeiträge.* Berlin: Universitätsverlag der TU Berlin.

Nobo, J. (1974). "Whitehead's Principle of Process". In: *Process Studies*, 4, (4, 1974), pp. 275-284.*

Pittendrigh, C. (1958). In: Roe, A.; Simpson, G. (eds.): *Behavior and Evolution.* New Haven: Yale University Press, pp. 390-406.

Prigogine, I.; Stengers, I. (1981). *Dialog mit der Natur: neue Wege naturwissenschaftlichen Denkens.* München: Piper. [(1984). *Order out of Chaos.* New York: Bantam Books.]

Proctor, J.; Larson, B. (2005). "Ecology, Complexity, and Metaphor". In: *BioScience*, 55, pp. 1065-1068. **

Schneider, E.; Kay, J. (1995). "Order from Disorder. The Thermodynamics of Complexity in Biology". In: Murphy, M. P.; O'Neill, L. A. J. (eds.): *What is Life: The next Fifty Years. Reflections on the Future of Biology.* Cambridge: Cambridge University Press, pp. 161-172.

Schrödinger, E. (1967). *What is life? – The physical aspect of the living cell.* Cambridge: Cambridge University Press.

Solé, R.; Goodwin, B. (2000). *Sings of Life. How Complexity pervades Biology.* New York: Basic Books.

Stegmüller, W. (1969). „Teleologie, Funktionanalyse und Selbstregulation". In: Stegmüller, W. (ed.): *Probleme und Resultate der Wissenschaftstheorie und Analytischen Philosophie.* Berlin: Wissenschaftliche Buchgesellschaft, pp. 639-773.

Toepfer, G. (2005). „Teleologie". In: Krohs, U.; Toepfer, G. (ed.): *Philosophie der Biologie.* Frankfurt am Main: Suhrkamp, pp. 36-52.

Whitehead, A. N. (1920). *Concept of Nature.* Cambridge: Cambridge University Press.

——(1929). The *Function of reason.* Princeton: Princeton University Press.

——(1967a). *Science and the modern world.* New York: Free Press.

——(1967b). *Adventure of Ideas.* New York: The Free Press.

——(1978). *Process and Reality.* New York: The Free Press.

——(1985). *Symbolism. Its Meaning and Effect.* New York: Fordham University Press.

* The articles from *Process Studies* are quoted in the online-version: www.religion-online.de.

** These articles are quoted from the Online-Version of the Journal *Bioscience*

The *Experience* of Environmental Phosphate Fluctuations by Cyanobacteria: an Essay on the Teleological Feature of Physiological Adaptation[1]

GERNOT FALKNER AND RENATE FALKNER

1. The insufficiencies of classical physics for understanding the relation between physiological adaptation and experience of environmental alterations

At the beginning of the 20th century physicists were confronted with certain experimental results that contradicted essential predictions of classical physics; they provided new insight into the behaviour of matter at the quantum level and revealed that at that level an observer is part of the observed quantum phenomenon. Hundred years later biology is in a similar situation: neurobiologists and physiologists begin to realize that a strict separation between an observed organism and the observer can no longer be maintained, when the interaction between organisms and their environment is studied. Difficulties arise from two problems: firstly, in this kind of studies the observed system is affected by the investigation, because the observer is part of the environmental influence. Secondly, physiologists and neurobiologists are confronted with phenomena that cannot be interpreted by a notion of substance from which everything associated with subjectivity, feelings, and intentions has been removed. This notion can be traced back to the Cartesian separation of reality into two entities, namely a material substance (*res extensa*) and a mental substance (*res cogitans*). The two substances differ in that the material substance is supposed to be composed of independently existing things, whereas the mental substance is

[1] This work has been supported by the Austrian Science Fund (FWF, Project Nr. P16237-B06).

considered to be the basis of conscious experience. In classical physics Newton's conception of absolute space and absolute time allowed localizing independently existing things, and to predict their motion when the forces acting upon them are known. The successful application of this conceptual scheme in physics has led to the idea that also biological processes can be explained by mechanistic models, in which localized biochemical reactions are assumed to be controlled by the regulatory influence of an omnipotent genome. This idea has dominated physiological research over the last decades, but turned out to be insufficient for investigating the vectored and irreversible nature of biological processes.

The insufficiencies of mechanistic models for interpreting biological activities are already revealed in experiments with lower organisms, in which their physiological adaptation to environmental alterations is studied. In these organisms physiological adaptation even takes place when the concentration of just one essential nutrient is altered in the growth medium. In this case a whole cascade of adaptive reconstructions sets in: in a first step the properties of the uptake system for the nutrient in question are altered, such that the system operates efficiently under the new condition. In a second step the whole cell transcends from one mode of organisation to another mode, in which the energy dependent uptake system and the overall energy conversion of the cell are conformed to each other. The resulting new type of cell then potentially grows at a different rate.

An experimental analysis of physiological adaptation confronts experimentalists and theoreticians with the problem that organisms adapt to the experimental conditions already during the course of the investigation, so that the outcome of the experiment depends on the way it is performed. This impairs an objective analysis of this phenomenon, a pre-condition for the establishment of a mechanistic model. An analysis of *adaptive events* using orthodox mechanistic models is also prevented by the fact that these models are confined to the Cartesian notion of material substance, in which only *external relationships* are assumed to exist between distinct portions of matter. Since one set of external relationship is as good as any other (Whitehead 1926/1985, 135), this notion of substance does not allow distinguishing between adapted and non adapted states of an organism. A reduction of the behaviour of a unicellular organism to an ensemble of interacting ions and molecules even does not allow differentiating between organisms and their environment, since molecules do not carry a label that indicates to which of the two regions they belong. Moreover, the Newtonian paradigm of classical physics holds another insufficiency for understanding life processes: during adaptive processes the energy and substrate flow of the microorganisms are adjusted to each other until a state of least energy dissipation is attained in the prevailing external milieu. As will be outlined below, this mutual adjustment of two or more cellular subsystems occurs simultaneously and hence in causal independence of each other. Since, in this process, the final adapted state affects the direction of the foregoing adaptive event, the irreversible flow of oriented adjustments along an axis of time has an inherent teleological character that is hard to understand within the framework of mechanistic interpretations, in which a cause always has to precede the effect.

For this reason an investigation of adaptive processes requires a theoretical foundation that is not restricted to basic postulates of classical physics. Such a foundation can be found in Whitehead's "philosophy of organism", in which the two Cartesian substances are transformed into the physical and the mental aspect of a process of self-creation. This idea can be transferred into physiology by relating the experience of environmental changes to the adaptive self-constitution of an organism, as originally proposed by John Dewey in 1925. The empirical investigations presented below are inspired by this idea. The obtained results provide new insights into biological self-organization. The fact that these results would not have been established without Whitehead's process philosophy sufficiently justifies a consideration of this philosophy in physiological research, dealing with analogies between experience and physiological adaptation. These analogies are reflected in the fact that every new experience proceeds in the context of a prehistory of experiences, being determined by events that are *learned* by the organism, because they have played a certain role in its development. At the physiological level this corresponds to the observation that an adaptive response is influenced by the outcome of preceding adaptations, so that in this respect an organismic memory is revealed in physiological processes. Furthermore, both experience and physiological adaptation are endowed with an *anticipatory* orientation towards the future, which is revealed in organismic *intentions* (as defined in section 6; see also: Muraca, this volume, sections 4, 5). Thereby, in every act of perception only a small portion of external factors is selected from a multitude of environmental influences, namely, those that can be coherently coordinated by the organism for an average best behaviour. In an analogous way, microorganisms adapt to a given environment by an appropriate coordination of uptake and signal transduction processes with the metabolism, such that the future requirements of the growing cell are met. Thus, both the experience of environmental changes and physiological adaptation has a temporal *vector character*, comprising the "history" of former influences and intentions for the future. Naturally this conception is incompatible with the Cartesian notion of material substance, because it refers to an organismic *self*, which is a prerequisite for maintenance of the difference between an organism and its environment. As will be shown below, in microorganisms this difference is dynamically preserved by adaptive processes that affect characteristic features of their energy converting subsystems.

1.1. The ontological difference between two manifestations of adaptive events

Using basic propositions of process philosophy, the temporal aspect of physiological adaptation can be treated by postulating that an *internal relatedness* exists between adaptive events. By internal relatedness we mean that the interdependence of all metabolic processes is altered when this organism adapts physiologically to an alteration of its ambient milieu. In this regard physiological adaptation can be considered as "a process of self-construction for the achievement of a unified experience" that "produces a new product, in which percepta in one mode and percepta in another mode, are synthesized into one subjective feeling" (Whitehead 1929/1978, 179; in this context Whitehead uses a generalized notion of experience from which all traits characteristic for higher animals only have been removed). The proposed relation between physiological adaptation and experience presupposes the existence of some sort of ontological difference between non-adapted and adapted states, since otherwise distinct adaptive events could not be distinguished from the constant flux of structural alterations of the cell. Adaptive events represent elementary "acts of becoming" (Whitehead 1929/1978, 69), by which an organism constitutes itself in a new experience of an environmental alteration, such that the resulting structure potentially serves the organism under the new ambient condition. In this regard "'becoming' is the transformation of incoherence to coherence, and in each particular instance ceases with this attainment" (Whitehead 1929/1978, 25). The ontological difference is revealed in two distinct manifestations of an adaptive event, namely *adaptive operation modes* and *adapted states*. An adaptive operation mode is initiated, when a previously attained adapted state is disturbed by an environmental alteration. In this mode a cell subjectively *interprets* (see section 5) an environmental alteration by a reconstruction of the pre-existing organismic structure. The result of this interpretation leads to a new adapted state, which then becomes an environmental factor for subsequent adaptive operation modes. The irreversible transition from adaptive operation modes to adapted states accounts for the teleological nature of an adaptive event. It is obvious that such teleological "acts of becoming" can only be found in dynamic manifestations of a cell and not in objectified aspects of an organism (such as a genome, proteome, etc.). This implies a consideration of metabolic flows and the energy conversion therein, involved in formation and degradation of the constituents of an organism.

Intracellular energy flow is accomplished by a great number of energy converting subsystems, such as the photosynthetic and respiratory electron transport chain in mitochondria, thylakoids or bacteria, in which the electron flow is coupled to a variable degree with the formation of ATP. ATP is the universal energy source for all kinds of energy consuming processes, each of them also representing an energy converting subsystem. Also the so called "two-component regulatory systems", monitoring *via* phosphorylation- and dephosphorylation-reactions external signals, can be considered as energy converting subsystems. By "cross-regulation" among different signal transduction pathways an information processing network can be established, coordinating the biosynthesis of different proteins with the subsequent growth requirement of a cell in a great many ways. Moreover, the cytoskeleton in growing and moving eukaryotic cells consists of interplay of many energy converting subsystems, potentially involved in information processing about environmental alterations during cell locomotion and morphogenesis. Last, but not least, a whole cell operates as one adaptive energy converting unit, when subcellular energy converters have been mutually adjusted to each other. In this process "the many become one and are increased by one" (Whitehead 1929/1978, 21).

Due to an inherent tendency to develop towards stationary states of least energy dissipation (Katchalski and Curran 1965), energy converting subsystems play a key role in the adaptive dynamics of microorganisms. Considering this energetic constraint, an adaptive event can now be defined as an occurrence, in which an energy converting subsystem passes from one energetically favourable adapted state to the next (Falkner and Falkner 2000). The constraint to operate with least energy dissipation has an important consequence: during fluctuations of the external substrate concentrations the cell must permanently make decisions about how to respond to a given fluctuation pattern, in order to select the concentration levels for which a reconstruction of its subsystems is useful in energetic terms. As a result the selected concentration potentially provides an average best operation of the reconstructed subsystem during future fluctuations. An adaptive operation mode, however, can only acquire that teleological character, when energy converting events have some sort of information processing capacities, contributing to the maintenance of the organism under ever changing environmental conditions. Thereby, adaptation to new environmental challenges must be guided by a cellular memory about foregoing adaptations, aimed to attain a new adapted state, in which the organism potentially is preserved (in this context we have employed a definition for information, originally given by Bateson (1972/2000); accordingly, an elementary unit of information is defined as a difference (in the environment) that makes a difference (in the organismic system)). The anticipatory feature of adaptive operation modes provides connectivity between subsequent adaptive events. Owing to the dependence of this mode on former influences, the adaptive response of energy converting subsystems can take place in many different ways. This leads to a multitude of organismic manifestations that have to be accounted for in subsequent adaptive events, such that excessive deviations from idealized organismic manifestations are avoided (see: Koutroufinis, this volume, Fig. 7 and 8).

In the following we present experimental evidence for this kind of information processing, occurring during external phosphate fluctuations. We give an example in which former adaptive events influence subsequent events, so that it depends on a historical concatenation of events, how an environmental alteration is experienced by the cells. The example deals with the complex interrelation between phosphate uptake and algal growth, exhibited in lakes under conditions under which the proliferation of phytoplankton is determined by the amount of inflowing phosphate.

2. Information processing by phytoplankton under natural growth conditions

Phosphate is an essential nutrient for growth of algae and cyanobacteria. In unpolluted, oligotrophic lakes these organisms are exposed to extreme fluctuations of the phosphate supply and the concentration of this nutrient frequently decreases to such low levels that the organisms do not have sufficient energy to drive the uptake process against the existing concentration gradient (Hudson et al. 2000; Falkner et al. 1989). However, the concentration can locally and for a short time increase to higher levels, for example after excretion of faeces by aquatic animals. In this situation algae and cyanobacteria have a special mechanism that allows exploiting efficiently transient increases in the ambient phosphate concentration. In periods, in which the external concentration exceeds a characteristic *threshold value* above which the available energy suffices to drive the transport into the cell, this nutrient is rapidly accumulated by an activated uptake system and then stored in polyphosphate granules. Simultaneously and because of the uptake activity of the whole community, the external concentration decreases to the threshold value and further uptake is only possible, when the external concentration rises again. As a result of this energetic constraint, fluctuations of the external concentration are experienced by the cells as *pulses*, in which transient increases of the external concentration are interrupted by periods without phosphate uptake. In continuous cultures the growth rate becomes greater with an increase in the amount of polyphosphates (Droop 1973).

By this mechanism phosphate uptake and growth are not directly coupled: phosphate uptake can occur in non-growing cells, and subsequent growth is possible without phosphate uptake at the expense of stored phosphate. In order to conform the growth rate to the amount of stored polyphosphates, both the properties of the uptake system and the growth rate have to be adjusted to the pattern of previously experienced phosphate pulses in an anticipatory manner, for the following reason: on the one hand, during short-term rises in the external phosphate concentration cyanobacteria must incorporate so much phosphate that in the subsequent period without phosphate supply growth is sustained at a rate, which – in turn – depends on the amount of stored phosphate. On the other hand, the polyphosphate granules must not become too large as to disrupt cellular structures. These two opposite constraints confront the cells with a regulatory problem, since the granules are segregated in the cytoplasm as osmotically inert structures, so that a direct effect of the size of the granules on the activity of the uptake system is not conceivable. To solve this regulatory problem, the cells need to memorize how much phosphate has been stored during previous phosphate accumulations and to coordinate during a pulse the adaptive properties of the uptake system with the growth rate accordingly (Falkner et al. 1995, 1996, 2006; Falkner and Falkner 2003; Wagner et al. 1995, 2000). This memory should be revealed by the adaptive behaviour in a sequence of phosphate pulses. Inspired by the ideas of Alfred North Whitehead, we may presume that the experience of phosphate pulses consists of a temporal nexus of adaptive events in which information about former events regulates subsequent events, such that the amount of accumulated phosphate meets the demand at the adjusted growth rate.

In the subsequent section we explain the molecular basis for the adaptive properties of the phosphate uptake system. We will then give a few examples of characteristic phenomena that can be observed when an interdependence of adaptive events is studied. Outgoing from these observations we will discuss the described adaptive behaviour within the framework of Whitehead's philosophy.

3. The adaptive properties of the phosphate uptake system

The incorporation of external phosphate into the polyphosphate pool proceeds in three steps: 1) Transport of external phosphate P_e into the cell. 2) Conversion of the incorporated internal phosphate P_i to ATP *via* photophosphorylation (in the light) or oxidative phosphorylation (in the dark). 3) Formation of polyphosphates from ATP. The adaptive features of this reaction sequence will be described below.

A "carrier protein" catalyzes the transport of phosphate from the external medium into the cell (reaction 1 in Fig. 1). At low external concentrations an energy source is needed for the translocation against the existing electrochemical gradient at the cell membrane. In the cyanobacterium *Anacystis nidulans* the necessary amount of energy is provided by an ATPase (Wagner and Falkner 1992), which can be coupled to the phosphate carrier to variable degrees (reaction 2 in Fig. 1). Since the ATPase receives its substrate from the ATP-synthase at the thylakoid membrane (reaction 3), the proton flux across the thylakoid membrane can be coupled with the transport process in a great many ways. In adapted states the degree of coupling is conformed to the external concentration such that energy dissipation is minimal. As a result of this energetic constraint the degree of coupling is increased, when the external concentration decreases. This affects the threshold value, which becomes lower, when the degree of coupling is increased (Falkner et al. 1993; 1994). Concomitantly with a decrease of the threshold value the activity of the transport system becomes higher (Falkner et al. 1989).

Also the next step, the conversion of internal phosphate into ATP (reaction 3 in Fig. 1), is energy dependent and potentially coupled to a variable degree to the driving process, which is the flow of protons from the thylakoid space into the cytoplasmic space. Furthermore, the H^+/ATP stoichiometry is variable; the higher the stoichiometry, the lower the stationary phosphate concentration in the cell. In adapted states the Michaelis constant of the ATP-synthase for phosphate is conformed to the steady state concentration, resulting from an adjustment of the H^+/ATP-stoichiometry to changing phosphate concentrations; this also provides efficient operation (Wagner and Falkner 1992). Only the last step, the transphosphorylation of the terminal phosphate group from ATP to polyphosphates (reaction 4 in Fig. 1) does not require an energy source. Reaction 5 stands for a passive loss of phosphate by the cell, as an example for a dissipative process and reaction 6 refers to photosynthetic CO_2-fixation. In the following treatment we designate with "transport system" and "ATP-synthase" all coherently functioning transport systems and ATP-synthases of the cell.

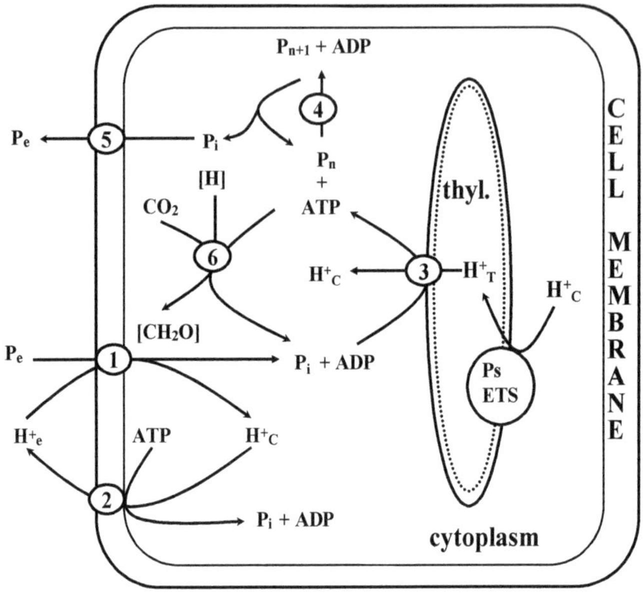

Fig. 1: Schematic presentation of phosphate utilization by cyanobacteria. The correct stoichiometries are not indicated in the figure. P_e: external phosphate; P_i: internal phosphate; P_n, P_{n+1}: polyphosphates; $[CH_2O]$: products of CO_2-fixation. H^+_e, H^+_C and H^+_T are the proton concentration in the external, cytoplasmic and thylakoid space. Ps ETS: photosynthetic electron transport system. For further details, see text.

By these adaptive accommodations the uptake system is tuned like a sensor to alterations of the external and cytoplasmic phosphate concentration. Any change in these concentrations initiates an adaptive interplay of the two subsystems that comes to an end, when the overall conversion of external phosphate to polyphosphates proceeds with optimal efficiency. In the resulting conformed state the dependence of uptake rate J_P on the external concentration $[P_e]$ obeys a simple linear flow-force relationship over a concentration range that extends from the threshold value to the external concentration, to which the system has been conformed. This relationship, which was originally proposed by Thellier in 1970, has the form:

$$J_P = L_P \times (\log[P_e] - \log[P_e]_A).$$

L_P is a conductivity coefficient that reflects the maximum velocity of the uptake system. $[P_e]_A$ is the threshold value of the corresponding adapted state. The extended range of validity of the linear dependence of the uptake flow on the logarithm of the external phosphate concentration can be explained by functional integration of high and low affinity transport systems during a *concrescence* to a conformed unity (Falkner et al. 1995, Wagner at al. 1995). It is notable that this linear flow-force relationship has the same structure as Weber-Fechner's law, when the uptake rate J_P is interpreted as the response to the stimulus $[P_e]$. Apparently in adapted states of cyanobacteria the relation between stimulus and response follows a similar logic structure as the sensory perception of higher organisms.

4. Experimental demonstration of microbial interpretations of changes in the external phosphate concentration

When growth of Cyanobacteria is limited by the amount of supplied phosphate, short-term rises of the external phosphate concentration above a threshold value lead to a variety of adaptive responses that can be interpreted as subjective "decision making", depending in a complex manner on preceding phosphate supply. In the following, we give characteristic examples of the capacity to interpret and memorize distinct patterns of phosphate fluctuations in an anticipatory manner. In the first example two differently diluted suspensions of the filamentous cyanobacterium *Anabaena variabilis* (corresponding to a total intracellular phosphorus content of 0.25 and 5.0 μMol per litre) were exposed to three phosphate pulses in order to analyze the time course of the decrease of the external phosphate concentration in respect to possible adaptive alterations of the uptake system during the uptake process. The amount of supplied phosphate in each of the three respective pulses was of the same magnitude as the original cellular phosphorus content of the two suspensions.

Fig. 2 shows that the suspension with the lower cell number per unit volume (upper graph) incorporated the supplied amounts of phosphate without significant adaptation, so that the time course of the decrease of the external concentration $[P_e]$ during the three subsequent pulses was practically identical. In contrast, the suspension with the twenty times higher number of cells

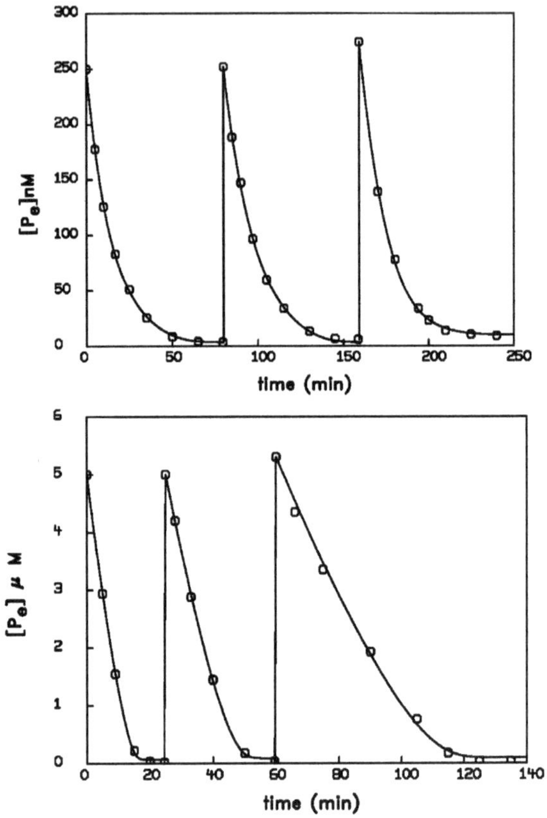

Fig. 2: Time course of ^{32}P-phosphate removal from the external medium by *Anabaena variabilis*. Two suspensions containing a cellular phosphorus content of 250 nMol/L and 5 µMol/L (upper and lower graph) were exposed to three consecutive pulses of phosphate, in which the population repeatedly received the same amount of phosphate as its original cellular phosphorus content. The curves represent the best computer fit using for the upper graph the equation (Thellier 1970): $J_P = L_P \times \log([P_e]/[P_e]_A) + L \times (\log([P_e]/[P_e]_A))^m$, with m = 5 for the first two pulses and m = 3 for the third pulse and for the lower graph the equation (Thellier 1970): $J_P = L_P \times \log([P_e]/[P_e]_A)$. For experimental details see: Falkner et al. 2006.

(Fig. 2, lower graph) revealed a totally different behaviour. In this case the kinetics of incorporation varied from pulse to pulse, indicating that an *adapted state*, attained by the uptake system during a foregoing pulse, influenced in a characteristic way the *adaptive operation mode* during the subsequent pulse. Apparently under this particular condition information about the prevailing adaptive response was propagated from one adaptive event to the next. Hence, although both populations had stored after the third pulse the same amount of phosphorus per average cell, the uptake system of the population with the lower number of cells still had its original activity, whereas the system of the population with the larger number of cells had been subjected to a considerable change.

The next example demonstrates the discriminatory potential of such kind of information transfer. In this case two identical samples of a population of *Anabaena v.* were exposed to one small pulse of 3 μM and one greater pulse with 6 μM, but in reverse order: one sample experienced first the small and then the great pulse, the other sample *vice versa*. A comparison of the uptake kinetics in a subsequent pulse of about 5 μM revealed significant differences (Fig. 3), indicating that the adaptive deactivation was lower, when the pulse height gradually increased in a sequence of pulses. The observed deactivation became greater with an increase in the number of antecedent pulses (Falkner et al. 2006).

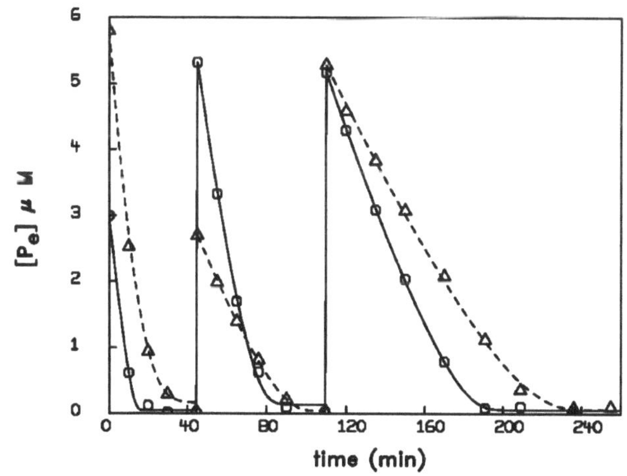

Fig. 3: Time course of ^{32}P-phosphate removal from the external medium by two identical population of *Anabaena variabilis*. The pulse pattern differs in the sequence of the first two pulses: one suspension was exposed first to the higher and then to the lower pulse (dashed line), the other *vice versa*.

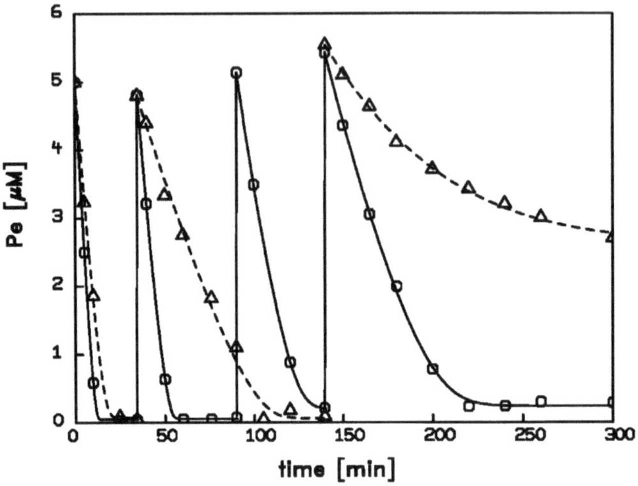

Fig. 4: Time course of ^{32}P-phosphate removal from the external medium by two identical populations of *Anabaena variabilis*, produced by different pretreatment of the same mother culture. Circles: "low pulse culture", triangle: high pulse culture.

We therefore may conclude that already a simple prokaryotic organism is capable of some sort of pattern recognition and to transform information about this pattern into a distinct adaptive response. Information about the pattern of experienced pulses can be transferred to daughter generations. When the two patterns of phosphate pulses, to which two different samples of the same population were exposed, differ significantly, a distinct adaptive behaviour could even be observed after subsequent cell division. In the following example we present an experiment in which an original culture of *Anabaena v.*, grown on a total phosphorus content of 10 μM, was diluted in two different ways, such that suspensions containing a cellular phosphorus content of 1.25 and 5.0 μMol were obtained. The two suspensions received seven times the amount of their original phosphorus content, but in two different supply modes: the suspension with the larger quantity of cells, designated as "high pulse culture", was exposed to seven higher pulses of 5 μM, whereas the more diluted suspension, termed "low pulse culture" experienced 33 lower pulses of 0.265 μM. After 24 hours of growth, sustained by the amount of stored phosphate, the two suspensions were brought back to the same cell density (corresponding to a total phosphorus content of 5 μMol L^{-1}). Both suspensions received a final phosphate quantity of 5 μMol/L, before they were then further cultivated under identical growth conditions for 48 hours at the expense of stored phosphate. After this growth period in which they doubled three times we investigated, whether the two different supply modes, experienced by the great-grandmother cells of the two compared cultures, led to different uptake behaviour in the great-grand daughter cells, when again challenged by a sequence of phosphate pulses, each at a concentration of 5 μM. Under these experimental conditions the two reference cultures showed similar uptake behaviour during the first pulse (Fig. 4). However, after the second pulse, the uptake kinetics differed considerably. Both cultures reduced the uptake rate, but much more so did the high pulse culture that experienced three cell divisions ago an elevated phosphate concentration. Still more pronounced became the difference during the next pulse. While the cells of the low pulse culture continued to incorporate phosphate in the third and even in a fourth pulse, uptake in the high pulse culture almost ceased at about three μM (extrapolated threshold value: 2.6 μM) after the third addition of phosphate. Hence, after 4 hours the low-pulse culture had stored ten times the amount of its original phosphorus content, whereas the phosphorus content of the high-pulse-culture increased about six times. It appears as if cells whose ancestors had experienced a high phosphate concentration in the past, *anticipated* by their behaviour a continuation of exuberant phosphate supply, when again exposed to elevated phosphate levels.

In the following we will refer the *subjective form* of interpretation of a particular sequence of phosphate pulses to the interplay of the two energy-converting subsystems, involved in the incorporation of external phosphate into the polyphosphate pool. A generalization of these properties to other adaptive energy converting subsystems of the cell shall then explain how a microorganism reconstitutes itself in every new experience of an environmental change. For this purpose the mutual adjustment of an ensemble of cellular energy converters during adaptation to changing phosphate concentration, corresponding to a *concrescence* of disjunctive *many* to a conjunctive unity of concrete togetherness (Whitehead 1929/1978, 21), will be analyzed into 5 factors: (i) the adaptive sensitivity of the uptake system in a particular environmental situation, resulting from antecedent challenges; (ii) the initial phosphate flow into the cell after the onset of a new pulse, leading to an intracellular disturbance; (iii) the interpretation of this disturbance by an adaptive operation mode in regard to a reconstruction of the uptake system; (iv) the phosphate flow after adaptation; (v) the intracellular tension (see below), resulting from the phosphate flow in the new adapted state (in this analysis we compared the reconstruction of the cell with the complex constitution of Whitehead's positive prehensions (Whitehead 1929/1978, 221).

5. Adaptive interplay of two energy converting subsystems during information processing about an environmental alteration

The experimental analysis presented in the previous section indicates that it is not possible to portray information processing by an objective model, since the experimental conditions also constitute information that is processed (in the examples given above it depends on the number of cells in the cyanobacterial suspension and the phosphate concentrations employed, to what extent information about former adaptations is transferred from one pulse to the next). Since in this case the experimentalist becomes part of the investigated system and of the organismic response to the experimentally imposed conditions, the experimental observation also has to be included in a model of information processing. However, in this case a modeller is confronted with the problem that also the observation of the observation had to be incorporated into the model, and so on, leading to some sort of "Self-Goedelization" (Koutroufinis 1996). Naturally this impedes an analysis of information processing in objectivistic terms. Furthermore, since every cell adapts to the concentration changes resulting from the uptake activity of the whole community, a dialectic relation exists between the cells of a community and the environmental alteration, caused by this community. In this dialectics the change of the ambient concentration and of the uptake system depend on each other, so that neither the change in the concentration nor in the system can be analyzed in terms of a simple cause-effect relationship: each transiently occupied adapted state is cause and effect simultaneously and the change of the concentration and of the uptake system reflect each other. It is notable that only under this condition an uptake system reveals its information processing capacity. In an attempt to give a non-mechanistic explanation of this phenomenon we will analyze adaptive events in respect to an analogy to Whitehead's "actual entities". From a formal point of view an analogy can be construed between interpretations, performed during adaptive events, and Whitehead's "prehensions" of actual entities, consisting in the present case of three factors (Whitehead 1929/1978, 23): (a) the adaptive manifestation of a certain cellular organization, referring to a subject that *interprets* a concentration change by a corresponding adaptive response; (b) an *interpreted object*, which is the pattern of phosphate pulses; (c) a *subjective form* of interpretation that determines the subsequent cellular self-constitution, leading to new properties of the phosphate uptake system and – in further processes – of the whole cell.

Developing this idea, it appears useful to distinguish two possible different types of adaptive behaviour. One corresponds to a situation in which adaptation to a pulse only transiently affects the transport system, so that the system reverts to its original state after the pulse; the other to a situation in which by adaptive interplay a conformed ensemble between the transport system and the ATP-synthase emerges. Due to a self-referential stability (see below) this ensemble survives then a period without phosphate supply and affects the uptake behaviour in the subsequent pulse in a distinct way. The former case refers – in a first approximation – to a temporal succession of adaptive events and shows a certain resemblance to Whitehead's temporal nexus of actual occasions of experience; the latter case applies to a temporal and spatial connectivity, corresponding to Whitehead's temporal and spatial grouping of actual entities (Whitehead 1933/1967, 201).

5.1. The temporal nexus of adaptive events

Let us discuss a situation, in which adaptation to the changing phosphate concentration is confined to the transport system. When in this situation a state of least energy dissipation, attained at the end of an antecedent pulse, is disturbed in a subsequent pulse, the transport system alters the degree of coupling between phosphate flow and ATP-hydrolysis in order to strive for another energetically favourable state. But as long as the concentration is further decreasing, no transiently acquired adapted state is stable. Therefore the uptake system will stepwise readapt to the gradually decreasing concentrations, until the lowest possible threshold level is attained (corresponding to the highest possible degree of coupling), where the system goes back to its original properties at the end of the foregoing pulse. In the subsequent pulse the same sequence of adaptive events is repeated, so that all pulses reveal similar kinetics, as shown in Fig. 2, upper graph. In this case the dependence of uptake flows on the driving force is non-linear and Weber-Fechner's law is therefore not valid over a wide concentration range. This non-linear dependence can be formalized, when the linear function given above is supplemented with non-linear terms (Thellier 1970). However, in accordance with the general line of reasoning about the connectivity of adaptive events, a phenomenological explanation can be given by postulating that the apparent non linear function is composed of many small linear sections, each of them representing a transiently attained energetically favourable state of the transport system (Thellier, personal communication). Since here the properties of the transport system are conformed to the momentary present concentration, they have to be rapidly replaced in a subsequent adaptive process, when the concentration further decreases; as a first approximation we may therefore consider this phenomenon as a temporal nexus of adaptive events that does not affect other energy converting subsystems of the cell. Naturally also in this situation every new adaptive adjustment would be influenced by the previously attained adapted state and to what extent this occurs, determines the shape of the overall non linear kinetics in a pulse.

5.2. Information processing and storage by a spatial and temporal grouping of adaptive events

A different situation arises, when during a pulse also the ATP-synthase and, in consequence, the overall energy conversion of the cell is conformed to transient increases of the cytoplasmic phosphate concentration during a pulse. In this case some sort of *communication* of the community of cellular energy converters takes place, by which the transport system and ATP-synthase are permanently reconstructed in a temporal and spatial nexus of adaptive events. By "communication" we mean a process in which the metabolic manifestations of the adapted states of the two systems are alternately *interpreted* in adaptive operation modes, so that adaptive alterations of the two systems influence each other. This communication, in which interpretations are performed in an anticipatory manner that potentially serves the organism, comes to an end, when a self-referential state between the two subsystems has been attained at the threshold value. In this state the kinetic properties of the transport system are conformed to the existing gradient between the stationary cytoplasmic phosphate concentration and the threshold concentration, determined by the energetic properties of the transport system. A similar adjustment seems to exist between the stationary cytoplasmic phosphate concentration, resulting from the energetic and kinetic properties of the ATP-synthase at the thylakoid membrane. Thereby, the Michaelis constant of this enzyme is tuned to the steady state concentration, determined by the respective H^+/ATP-stoichiometry (Wagner and Falkner 1992). Naturally these adjustments must be conformed to each other such that growth is sustained at a rate which is anticipated in the light of previous exposures (Wagner et al. 2000).

The anticipatory features of this communication, in which processed information depends on how cellular interpretation of the changing external concentration affects the properties of the two subsystems, can be graphically portrayed by a "family tree of individual occasions of experience" (adopted from Hansen 2004, Fig. 5). Each arrowhead is a symbol for an adaptive event in which the degrees of coupling between phosphate flow and ATP hydrolysis on the one hand and ATP production and proton flow on the other hand (see Fig. 1) are altered from q_n and Q_n to q_{n+1} and Q_{n+1} respectively. The lines between the arrowheads stand for external (dashed lines) and internal (full lines) influences, resulting from the outcome of antecedent adaptive events. They represent efficient causes for the onset of the subsequent adaptive processes. Their anticipatory interpretation, however, leading to the creation of a new cellular constituent which is supposed to provide an integer performance of the whole organism, requires explanation by some sort of "final cause" that accounts for the teleological feature of physiological adaptation. The temporal transition from potentiality *via* actuality into pastness, from where it becomes a causal factor for a new adaptive event, is the ultimate foundation for the irreversibility of the life process and gives rise to the historical dimension of adaptive events: when adaptations are guided by interpretations that are derived from the outcome of former adaptations, these former adaptations also represent interpretations of the more remote past, and so on. Hence, even for an understanding of the functioning of a unicellular organism, two levels of descriptions are required: one refers to its objective manifestation in adapted states; the other to the *decision making* in an adaptive operation mode (Plaetzer et al. 2005). Due to the vast number of possible stable tertiary structures of a protein the components of energy converting subsystems can be conformed to each other in a great many ways, thereby establishing functioning-dependent structures (Thellier et al. 2004) with many different kinetic and energetic properties. For this reason every adaptive event is a "act of becoming" of something new and every adapted state is an emergence of novelty: an attribution of a persistent structure to an organism or a species is based on arbitrary abstractions.

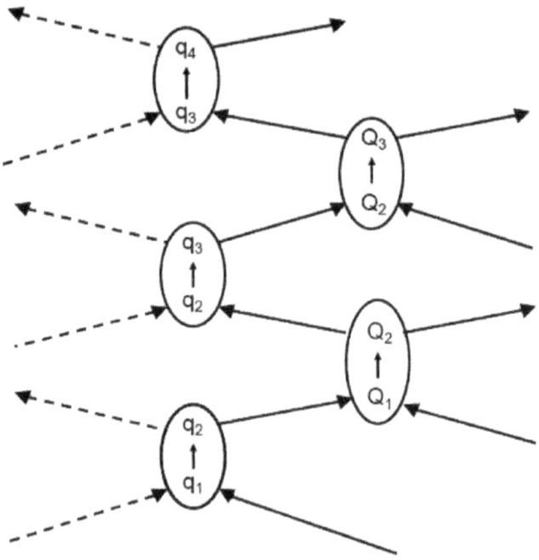

Fig 5: Family tree of interrelated adaptive events (adopted from Hansen 2004).

The self-referential state of the two subsystems, finally attained at the threshold value, provides stability, so that its adapted state is preserved during a subsequent growth period without phosphate supply and even appears to survive a possible turnover of proteins. This explains how a distinct pattern of former pulses influences a subsequent pulse in a specific way. When one of the proteins of the uptake system is occasionally replaced in the cellular turnover of proteins, the new protein will readapt to the preexisting condition and, by this mechanism, "inherits" the properties of the substituted protein. The phosphorylation potential resulting from a self-referential ensemble of transport system and ATP-synthase then determines the subsequent reconstruction of the cell in regard to the anticipated growth rate (Plaetzer et al. 2005). In this process a multitude of translational and post-translational modifications are directly or indirectly affected. In order to operate at the anticipated growth rate with optimal efficiency, other cellular energy-consuming subsystems also have to accommodate to the energetic and kinetic properties of the ATP-synthase with respect to least energy dissipation (Stucki 1980). This implies that also the other energy consuming subsystems of the cell are altered in adaptive operation modes in which the cell also *subjectively interprets* the objective manifestations of the uptake system in its stable self-referential state. Since a new energetic situation influences all energy-dependent subsystems *simultaneously*, an adjustment of other energy consuming subsystems to the energetic manifestations of the ATP-synthase can be expected to occur in a coherent manner, until a new adapted state of the whole cell emerges.

6. The permanent reconstitution of the cell

The question has to be answered how a characteristic form of the cell is maintained during a constant flux of structural rearrangements of all cellular subsystems. We have outlined above that adaptive operation modes, directed towards attainment of a conformed adapted state, cannot be analyzed in terms of spatially extended entities, existing independently of each other, because in this mode both the involved constituents and the cellular environment are in flux. An experimental localization of participating components disturbs this flux and alters its direction. The same refers to the concrescence of the overall energy flow through the cell towards a state of least energy dissipation, consisting of contemporary events that happen in causal independence of each other (Whitehead 1923/1978, 61). A coordinating principle is therefore needed in order to explain how simultaneously occurring events finally lead to an organismic unity of mutually adjusted subsystems.

In reference to Whitehead's ideas about the nature of actual entities a possible explanation can be given in that an adaptive reconstruction of the cell is dominated by a "subjective aim" by reason of which a cell decides to what extend an external change has significance for itself. An actual entity determines itself by combining "self-identity" with "self-diversity" (Whitehead 1929/1978, 25), by analogy with an organism that maintains its identity, although its constituents are permanently replaced. In order to obtain a physiological basis for this "subjective aim", we may postulate that the energy flow through a living system serves as a medium for a field of tension, guiding anticipatory information processing about a potentially useful reconstruction of energy converting subsystems in adaptive operation modes. This field is reflected in an "internal relatedness" among "acts of becoming", which is a precondition for the "*conceptual feeling*" of an "*ideal of itself*" (Whitehead 1929/1978, 87). Without such an "ideal of itself", corresponding to a complex "eternal object" (in Whitehead's terminology), maintenance of a certain form of a (lower) organism in a constant flux of metabolic conversions is difficult to explain. This proposition rejects the orthodox opinion that microorganisms owe invariant features of their form unchangeable structural components (such as DNA-sequences that are supposed to function as genes). It is replaced by the idea that self-identity of an organism is provided by a "defining characteristic" (Whitehead 1933/1967, 204), accounting for a connectivity of adaptive events in every act of experience. Any model of "self organization" that does not combine "self-identity" with "self-diversity" portrays a "self organization without self" (Koutroufinis 1996).

According to this concept the "subjective aim" to achieve a tension-free state manifests itself as *intentions* that guide as *final cause* the concrescence of different cellular flows towards a single, even though complex organismic entity. Naturally such a tension-free state only represents an idealized model of itself, in the sense of Rosen's (1985) postulate for anticipatory systems. During growth it is never really attained in a defined cellular structure, since under this condition the ratio of substrates required for biosynthesis of cellular constituents is not invariant and therefore the flows of nutrients into the cells have to be constantly re-coordinated. For this reason the mutual adjustment of subsystems in regard to efficient operation can be expected to be rhythmic, oscillating between more or less conformed states. In this case also energy converting subsystems in the cell membrane will be included in this rhythmic process. From the adaptive interplay of ATPases, channels and transport systems in the cell membrane also the ionic milieu in the extracellular space, in which cell wall biosynthesis takes place, is locally affected. Since the cell wall determines the shape of a plant cell, the rhythmic formation of conformed adapted states potentially has an important implication for morphogenesis. Thus, a certain form is maintained during growth by rhythmic transitions from initial flows to conformed adapted flows, accomplished by organism-specific interplay between efficient causes and final causes. Efficient causes are given on a cellular level by the respective potential difference between the substrates and products and the catalytic properties of the involved enzyme systems that catalyze the corresponding reactions in adapted states. Final causes are guided by a "conceptual feeling of subjective aim" (Whitehead 1929/1978, 224), controlling the network of adaptive operation modes, in which previously attained adapted states are altered in regard to anticipated flows that preserve the shape of the organism.

References

Bateson, G. (1972/2000). *Steps to an Ecology of Mind.* Chicago: The University of Chicago Press.

Dewey, J. (1925/1958): *Experience and Nature*. New York: Dover Publication.

Droop, M. (1973). "Some thoughts on nutrient limitation of algae". In: *J. Phycol.*, 9, pp. 264-272.

Falkner, R.; Priewasser, M.; Falkner, G. (2006). „Information processing by Cyanobacteria during adaptation to environmental phosphate fluctuations". In: *Plant Signaling and Behaviour*, 1, pp. 212-220.

Falkner, R.; Falkner, G. (2003). "Distinct adaptivity during phosphate uptake by the cyanobacterium Anabaena variabilis reflects information processing about preceding phosphate supply". In: *Journal of Trace and Microprobe Techniques*, 21, pp. 363-375.

Falkner, G.; Falkner, R. (2000). "Objectivistic views in biology: an obstacle to our understanding of self-organisation processes in aquatic ecosystems". In: *Freshwater Biology*, 44, pp. 553-559.

Falkner, G.; Wagner, F.; Falkner, R. (1996). "The bioenergetic coordination of a complex biological system is revealed by its adaptation to changing environmental conditions". In: *Acta Biotheoretica*, 44, pp. 283-299.

Falkner, G.; Wagner, F.; Small, J.; Falkner, R. (1995). "Influence of fluctuating phosphate supply on the regulation of phosphate uptake by the blue-green alga Anacystis nidulans". In: *J. Phycol.*, 31, pp. 745-753.

Falkner, G.; Wagner, F.; Falkner, R. (1994). "On the relation between phosphate uptake and growth of the cyanobacterium Anacystis nidulans". C. R. Acad. Sci. Paris, *Sciences de la vie/Life sciences*, 317, pp. 535-541.

Falkner, G.; Falkner, R.; Wagner, F. (1993). "Adaptive phosphate uptake behaviour of the cyanobacterium Anacystis nidulans: analysis by a proportional flow-force relation". In: C. R. Acad. Sci. Paris, *Sciences de la vie/Life sciences*, 316, pp.784-787.

Falkner, G.; Falkner, R.; Schwab, A. (1989). "Bioenergetic characterization of transient state phosphate uptake by the cyanobacterium Anacystis nidulans. Theoretical and experimental basis for a sensory mechanism adapting to varying environmental phosphate levels". In: *Arch Microbiol*, 152, pp. 353-361.

Hansen, N. (2004). "Spacetime and Becoming: Overcoming the contradiction between special relativity and the passage of time". In: Eastman, T. and Keeton, H. (Eds.). *Physics and Whitehead*. New York: State University of New York Press.

Hudson, J.; Taylor, W.; Schindler, D. (2000). "Phosphate concentrations in lakes". In: *Nature*, 406, pp. 54-56.

Katchalski, A.; Curran, P. (1965). *Nonequilibrium Thermodynamics in Biophysics*. Cambridge: Harvard University Press.

Koutroufinis, S. (1996). *Selbstorganisation ohne Selbst. Irrtümer gegenwärtiger evolutionärer Systemtheorien*. Berlin: Pharus Verlag.

Plaetzer, K.; Thomas, S. R.; Falkner, R.; Falkner, G. (2005). "The microbial experience of environmental phosphate fluctuations. An essay on the possibility of putting intentions into cell biochemistry". In: *J. Theor. Biol.*, 235, pp. 540-554.

Rosen, R. (1985). *Anticipatory Systems. Philosophical, Mathematical and Methodological Foundations*. Oxford: Pergamon Press.

Stucki, J. (1980). "The optimal efficiency and the economic degrees of coupling of oxidative phosphorylation". In: *European Journal of Biochemistry*, 109, pp. 269-283.

Thellier, M. (1970). "An electrokinetic interpretation of the functioning of biological systems and its application to the study of mineral salts absorption". In: *Annals of Botany*, 34, pp. 983-1009.

Thellier, M.; Legent, G.; Norris, V.; Baron, C.; Ripoll, C. (2004). "Introduction to the concept of functioning-dependent structures in living cells". In: C. R. Acad. Sci. Paris, *Sciences de la vie/Life sciences*, 327, pp. 1017-1024.

Wagner, F.; Sahan, E.; Falkner, G. (2000). "The establishment of coherent phosphate uptake behaviour by the cyanobacterium Anacystis nidulans". In: *Eur. J. Phycol.*, 35, pp. 243-253.

Wagner, F.; Falkner, R.; Falkner, G. (1995). "Information about previous phosphate fluctuations is stored via an adaptive response of the high-affinity phosphate uptake system of the cyanobacterium Anacystis nidulans". In: *Planta*, 197, pp. 147-155.

Wagner, F.; Falkner, G. (1992). "Concomitant changes in phosphate uptake and photophosphorylation in the blue-green alga Anacystis nidulans during adaptation to phosphate deficiency". In: *J. Plant Physiol.*, 140, pp. 163-167.

Whitehead, A. N. (1933/1967). *Adventures of Ideas*. London: The Free Press.

—— (1929/1978). *Process and Reality*. New York: The Free Press.

—— (1926/1985). *Science and the Modern World*. London: Free Association Books.

Beyond Systems Theoretical Explanations of an Organism's Becoming: A Process Philosophical Approach[1]

SPYRIDON A. KOUTROUFINIS

This essay may be read as an application of one of the most central ideas in Whitehead's work: that the main task of metaphysical schemes of thought is to *criticize scientific abstractions* or, more specifically, *to criticize the confusion of something abstract with something concrete* in different sciences.[2] This will be discussed in the context of a classical topic of natural philosophy, the *equifinality* of an organism's generation. With this term Ludwig von Bertalanffy, founder of formal bio-systemic thought in the early 20th century, referred to the essential tendency of living beings, and open systems in general, to reach a certain terminal state via different possible developmental pathways. According to von Bertalanffy, the central subject of theoretical biology is the problem of *ontogenesis*, i.e., the generation of an adult multicellular organism. This is a question which has not adequately been considered by philosophers in recent decades. Instead, they focused their attention on supposedly more interesting issues, such as the evolution of the species, the generation of life and above all the nature of consciousness. Unfortunately, the choice of focus ignored the fact that the metabolism of even the simplest bacterium is more than just a very complex physicochemical system. It was not for nothing that Aristotle had made the problem of the self-sustainability of a living being and of its embryonic development the main subject of biophilosophy: a position it held until the rise of Darwin's theory of evolution.

Since the 1930s, most bioscientists have proceeded from the assumption that organisms arise and preserve themselves by means of efficient

[1] I gratefully acknowledge the editorial help and critical remarks of Terrence Deacon, Robert Valenza, and Andrew Packard.

[2] See Whitehead 1979, 7f.; 1953, 70.

causation and that only blind forces such as those studied by physics and chemistry, are at work in organisms. In modern biology there seems to be no place for a kind of thinking which I will later describe as *mentalistic teleological thinking.*

This essay is divided into two parts. The aim of the first part is to show that thinking embryogenesis only in terms of efficient causation, which operates on the basis of the theory of nonlinear dynamical systems, poses serious problems. Failing to recognize this would be a clear case of Whitehead's "fallacy of misplaced concreteness", meaning the confusion of the abstract with the concrete. In an attempt to overcome this problem, the second part of the article presents an alternative approach to teleology which I call *mentalistic teleology.*

1. The insufficient understanding of teleology in current bio-systemism

In the first half of the 20th century the attempt was made to banish all teleological thinking from biology. While biologists and philosophers of biology still talk about "teleology", it is not always clear what they mean by this term. Difficulties of definition apart, it is possible to describe teleology as the theory of events which tend to reach certain terminal states, to stay in them or to oscillate around them.[3] The definition is metaphysically neutral. It leaves open the question of whether the causes prevailing in organisms are mental or deterministic-material. Therefore, it is applicable both to living and lifeless nature and even to mechanical devices such as robots that are controlled by programs.

Most biologists are similarly content to use teleology as a synonym for equifinality. In contemporary biology the term "teleology" is used in connection with organisms, organs and behaviors to express the idea that these all display an affinity to a particular terminal state *strictly because of their own scientifically measurable efficient causality* (see below). This separates scientific teleology from the ancient Greek concept of teleology which was based on the metaphysical idea of final causality – Aristotle's *telos* or *causa*

[3] For similar positions see Mayr 1991, 59; 2000, 416, 405; Toepfer 2005, 36; Hull 1974, 103ff.

finalis – and the presupposition that an organism is shaped by *(proto)mental factors* acting within it.[4] Biophilosophy as distinct from philosophy of biology (see introduction of this volume) is free to introduce new perspectives on teleology going beyond modern biosciences.

1.1. On the current scientific concept of organisms as complex dynamic systems

A system is defined as a *dynamic system* if its state at any given moment can be described as a limited set of time-dependent or state variables $x(t) = [x_1(t), x_2(t), ..., x_n(t)]$ for which a function F can be formulated stating mathematically the connection between states at times t and $t + \delta t$. The properties of this function reflect the *causal* relationships at work within the system. The set of state variables $[x_1(t), x_2(t), ..., x_n(t)]$ spans an *abstract space*, the system's so-called "state-space".

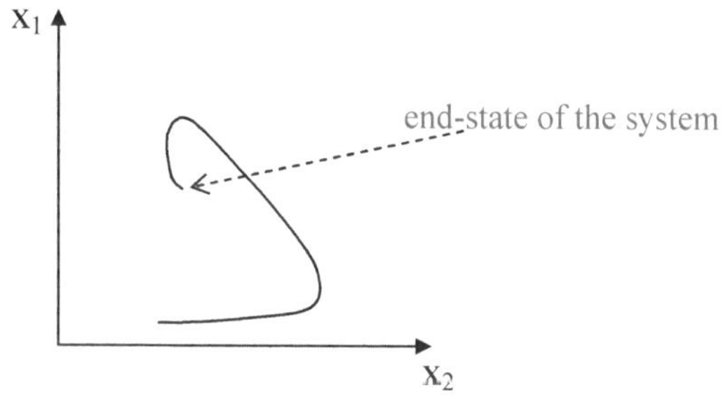

Fig. 1: State-space and trajectory of a dynamic system with two state variables.

It is important to keep in mind that the change or development of a dynamic system is not merely the result of the function F, but depends also on a group of externally fixed parameters. The most abstract formula for a dynamic system must therefore be (Ebeling and Sokolov 2005, 40):

$$x(t + \delta t) = F(x(t), p, \delta t);\ p = [p_1, p_2, ..., p_m] \qquad \text{(formula 1)}$$

[4] For more details about the history of the concept of teleology from antiquity to 20th century see Koutroufinis 2013, 2012.

The letter p represents a set of *parameters*. All parameters are externally fixed constants. They represent either real constants or quantities assumed as being constant, the latter being the usual case. *Their role is to constrain the development of the state variables x(t).* Usually, every state in a stable dynamical system can be calculated from its preceding state. This is sometimes not possible in unstable systems because some states have more than one possible successor state under actual (natural) conditions (see fig. 5). But even these systems allow, at least in principle, the calculation of all *potential* future states of the system's development given its state at a certain time. The fact that every potential state in a given dynamic system, whether stable or unstable, can be mathematically calculated from its preceding state shows that a dynamic system is governed by *efficient causality*. By "efficient causality" I mean the following:

- Firstly, that the state of a system at any given time is exclusively the function of its state and the state of the internal and external factors at work on it at the immediately preceding point in time.
- Secondly, that the transition from one state to the next is governed entirely by factors – whether intrinsic to the system or from its environment – which can be completely described in modern scientific terms,[5] especially in terms of today's physics and chemistry (with the exception of quantum theory, which is sometimes understood in terms associated with human consciousness). By definition, these descriptions exclude mental aim-oriented final causes and thus any non-scientific teleology.

It follows that the transition can be *exhaustively* plotted in abstract spaces designed on the basis of classical physics[6] and chemistry. So, all natural events whose actual (fig. 1) or potential (fig. 5 and 7) development can be displayed by trajectories in abstract spaces and whose causality can be ex-

[5] Such factors are: firstly, the *laws of physics and chemistry*; secondly, *quantities in physics and chemistry*, appearing either as dynamic variables or as constants (i.e., parameters; see section 1.3), such as the concentration of substance X, its reaction speed, pH value, pressure, temperature, free energy, etc.; and, thirdly, *stochastic* factors influencing the outcome of real processes in experimental physics and chemistry, such as thermodynamical and quantum-physical fluctuations.

[6] The term "classical physics" includes theories of self-organization or complexity based on non-linear dynamic systems' theory.

haustively reduced to scientifically describable factors (see. footnote 5) are merely governed by efficient causes. Dynamic systems can be subdivided into conservative and dissipative dynamic systems. The energy of the latter "dissipates", that is, it disperses and must be replaced by the environment. *Dissipative systems produce entropy.* As we shall see, it is precisely the fact that they produce entropy that allows dissipative systems (under certain conditions) the kind of spontaneous structuring of their behavior in space-time commonly labeled as "self-organization".

1.2. The paradox of self-organization: system organization spontaneously increases in the process of destroying the cause of its increased organization

Statistical entropy is a concept applicable only to systems with a huge number of particles and serves as a measure of disorder. Boltzmann and Planck define the statistical entropy of a system as the average value of its uncertainty (Ebeling 1976, 13).[7] A system is uncertain if there are many possible states in which it might be. Ideally, each of these states corresponds to a point in its state-space which can, with a particular probability, be the actual state of the system. The statistical entropy of a system is the average value of the probabilities of all possible states which the system might occupy. Accordingly, statistical entropy is related to the concept of *possibility*.

The order and the statistical entropy of a system which has n degrees of freedom can be depicted using a high-dimensional state-space (fig. 2).[8] For

[7] Concerning the connection of entropy and uncertainty see Ebeling and Sokolov 2005, 85f.

[8] The number of degrees of freedom that a system has depends on both its constitution and the way in which it is described. In statistical mechanics, 6m degrees of freedom are assigned to a system which consists of m elements. This is the case because each element has 6 degrees of freedom. Its complete description requires three spatial positions: its location in the three spatial dimensions x, y and z; and the three velocities v_x, v_y and v_z with which it moves along these dimensions. Chemical systems consisting of n different molecular species (the concentration of which can vary) are considered to be completely described if the concentration of all n molecular species is indicated at each point in time (see fig. 7). Thus they have n degrees of freedom.

each of the possible states there is a particular point in this abstract space to which it corresponds. Thus a limited area of the state-space represents all the states which the system can possibly occupy at a given time.

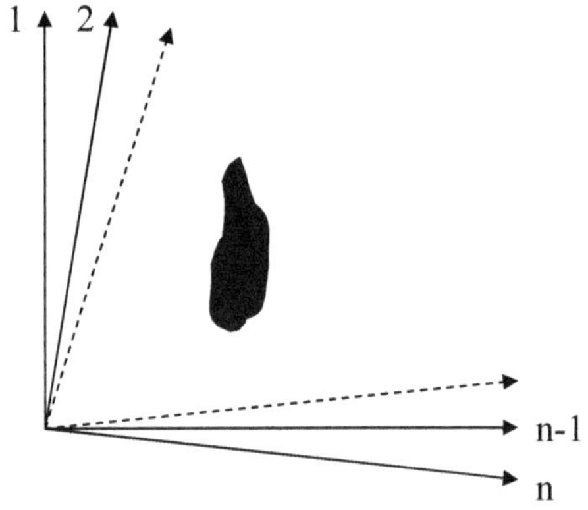

Fig. 2: The high-dimensional state-space-volume of the possible states of a system with n degrees of freedom at a given time.

The conviction that living beings are self-organized complex dynamical systems is central to current theoretical biology. "Self-organization" is a technical term. It means that the increase of a system's order – that is, the decrease of its entropy – is the result of efficient-causal interactions between its elements and not the outcome of the action of a single real or ideal entity such as an acting person or a program. Self-organization does not mean elimination of entropy or uncertainty but just their diminution (fig. 3).

Systems serving as models of self-organization require *gradients* of energy and/or material. A typical example of such a gradient is the difference of temperature T_1 - T_2 in the so-called Bénard convection. This occurs when the lower layer of a fluid is heated and the upper layer is kept at a cooler temperature (T_2). At a certain difference of temperature between the bottom and the top of the fluid, the heat flux reaches a critical value and convection arises. Coherent macroscopic movements emerge in the fluid and form a highly structured pattern of hexagonal cells (fig. 4). This phenomenon represents a so-called "phase transition", i.e. the change of one mode of dynamics to another: in this case the transition from heat conduction to convection. In Bénard convection the phase-transition, i.e., the self-organized pattern formation, does not occur immediately but takes several minutes (depending on the fluid).

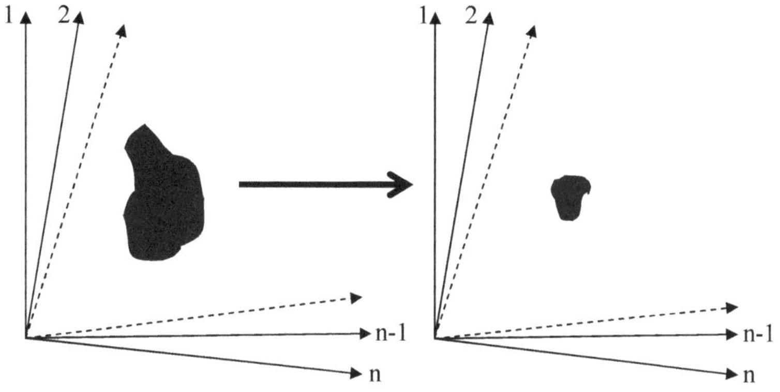

Fig. 3: The diminution of a system's entropy equals a decrease in the number of possible states.

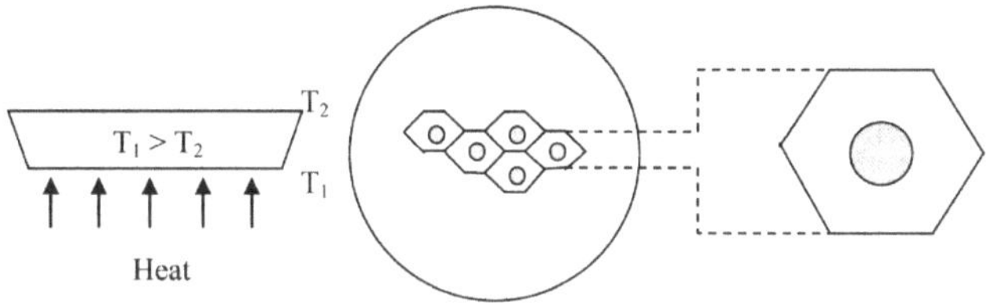

Fig. 4: The Bénard convection. The upward movement of the heated fluid in the center of each cell forms a cylinder. Fluid that has cooled down flows downwards in the hexagonal area around the cylinder.

There is a fundamental finding in thermodynamics with consequences for the applicability of the theory of complex dynamical systems to biology; these consequences are often misunderstood. It states that every form of self-organization of a physico-chemical system amounts to a *decreasing* of the gradients which are imposed on the system and which move it away from the thermodynamic equilibrium, that is, from the state of total lack of physical becoming. *Each self-organized system tends to return to equilibrium.* The hexagonal Bénard cells transport heat upwards faster than simple heat conduction thus *increasing the rate of gradient destruction.* Two well-known physicists go to the heart of this finding:

> "As systems are removed from equilibrium, they will utilize all avenues available to counter the applied gradients. As the applied gradients increase, so does the system's ability to oppose further movement from equilibrium. [...] No longer is the emergence of coherent self-organizing structures a surprise, but rather it is an expected response of a system as it attempts to resist and dissipate externally applied gradients which would move the system away from equilibrium" (Schneider and Kay 1995, 165).

According to this position, *all self-organized phenomena that occur in inorganic systems arise only because they oppose the causes of their self-organization.* The decrease of entropy inside a system enables it to increase the rate of degradation of the externally imposed gradients.

This essential property of dissipative dynamic systems means that they increase their *entropy production* in the process of decreasing of their own entropy. This is not paradoxical, since entropy and entropy production are two different quantities.[9] Open systems may increase their entropy production and at the same time decrease their local entropy if they export entropy faster than they produce it.

It is easy to see why self-organized pattern formation involves both a tendency to oppose displacement away from equilibrium and an increase of entropy production. Firstly, work is required to resist externally imposed factors that tend to displace the system away from equilibrium. Secondly, work can only be performed by a system that in some way constrains the flow of energy through it, i.e. an ordered system. In other words, self-organization involves the *intensification* of a system's mutually supporting (i.e. correlated) internal processes. Thirdly, according to the second law of thermodynamics all real (and not idealized) physical processes produce entropy. Therefore intensification of the correlated dynamics within a self-organized system increases its capacity to do work to oppose displacement from equilibrium *and* increases the rate of entropy production. In Bénard convection, for example, the gradual formation of the hexagonal cells enables the system to produce more entropy than it did before the phase transition from heat conduction to convection took place.

---

[9] In physics the symbol for entropy is S and for entropy production P. Entropy production is a rate. It has the dimension "entropy divided by time".

Production of entropy means the dissipation of energy or degradation of the capacity to do work. Energy cannot be created or destroyed in a system (forbidden by the first law of thermodynamics or law of conservation of energy) and the system cannot upgrade already degraded energy (forbidden by the second law of thermodynamics) in order to degrade it again. Thus the system can produce entropy as dictated by the second law only if it degrades the energy which it is supplied with from the outside, that is to say by the externally applied gradients. In other words, in order to oppose external gradients the system does work that uses (degrades) the energy provided by these same gradients.

So from the point of view of thermodynamics the decrease of a system's local entropy, i.e., its self-organization, requires an increase of its entropy production. This does not constitute an extremum principle of physics as is often characterized in the "principle of maximum entropy production."[10] This is because the increase of entropy production may stop before this quantity reaches its possible maximum.[11]

To conclude, the emergence of self-organized higher-order macroscopic structure (pattern formation) in inorganic dissipative systems, *only serves the degradation of externally imposed gradients through entropy production*. It is therefore ultimately self-destructive.

1.3. Examples of modeling biomolecular processes in systems biology

Some bioscientists, primarily systems biologists, maintain that organisms *are nothing more* than dynamical systems that can be expressed using the language and terminology of physics and chemistry. They eagerly anticipate the development of mainframe computers capable of providing computer-simulations of whole organisms, which, they assume, will be possible within the next fifty years.[12] In doing so, they attribute ontological and not just heuristic relevance to the theory of dynamical systems, making any such development philosophically interesting.

---

[10] See Sagan 2008, Salthe 2010, Swenson 1997.

[11] For criticism of the principle of maximum entropy production see Nicolis and Nicolis 2010, Ross et al. 2012.

[12] See Wolpert 1995, Tomita 2001, Normile 1999, Wayt Gibbs 2001.

Both the solving of nonlinear differential equations and concomitant computer simulations are fundamentally important for formal reductions of cellular processes in contemporary systems biology. But both operations require some specific conditions. I will focus on a currently inescapable methodical reality which is of major importance for the philosophical discussion about the relevance of self-organized dynamic systems theory to biology: the sharp distinction between dynamic and static quantities or, in other words, between variables and parameters (see formula 1).

Parameters may represent specific quantities or they may be *abstractions* that summarize the relations between quantities describing cell properties (such as volume, temperature, pressure, pH-value, etc.). In simulations, parameters are assigned by systems biologists. They are either experimentally derived, estimated or simply taken from other published studies. The sharp distinction between variables and parameters is readily apparent in many texts on systems biology. In the following, three typical examples will be briefly introduced.

The first example refers to bistable behavior, which is a special kind of instability. Bistability is a phenomenon often encountered in the theory of dynamic systems. Bistable systems are common in systems biology. Bistability is philosophically interesting as it shows that there are areas of indetermination in the development of some dynamic systems. Gardner et al. published in *Nature* a model for the mutual regulation of the activity of two genes which was developed on the basis of the well known Operon-model of Jakob and Monod. Both genes transcribe a protein – a so-called repressor protein – which blocks the activity of the other gene, so that both genes inhibit each other. The dynamics of this system of two interwoven *negative feedback* relationships can be described by two state variables, U and V, which are associated with the concentrations of both repressor proteins. The variation of the concentrations of both proteins can be represented by two differential equations (Gardner et al. 2000, 339).

$$\frac{dU}{d\tau} = \frac{\alpha_1}{1+V^{\beta}} - U$$

$$\frac{dV}{d\tau} = \frac{\alpha_2}{1+U^{\gamma}} - V$$

These coupled non-linear equations are so-called "dimensionless" equations.[13] The quantities α_1, α_2, β and γ are the *parameters*. Their value is determined by the experimenters. They are, however, kept constant in each single experiment and corresponding computer simulation as well. There are many ways to manipulate the values of the parameters – one possibility is through the variation of the ambient air temperature. As for the suitability of dynamical systems theory to explain the dynamics of an organism, one must first and foremost keep in mind that even the modeling of the self-organization of just two dynamic quantities requires that four static quantities or parameters be externally determined.

Certain combinations of the four parameters lead to a bistable behavior. Such a system has two possible stable terminal states. The corresponding state-space, with some of the potential trajectories, would then appear as follows:

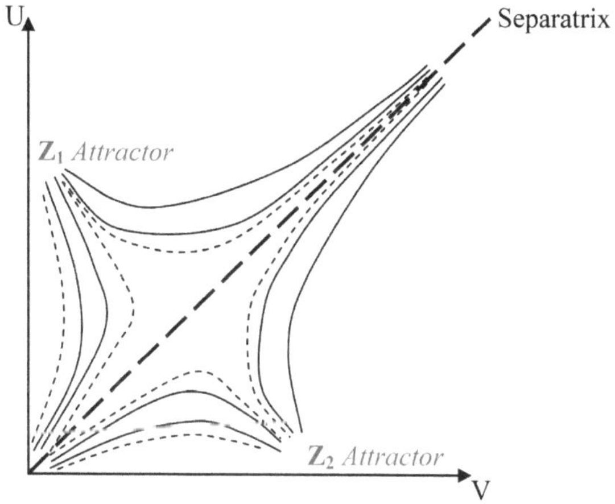

Fig. 5 (Based on Gardner et al. 2000, 340 and Ebeling et al. 1990, 150): For certain adjustments of the parameters a bistable behavior emerges, meaning that two stable alternative terminal states, Z_1 and Z_2, are possible.

[13] In dimensionless equations the quantities have to be recast by judicious scaling so that no units need appear. In the two equations provided by Gardner et al. the left sides seem to have the quality of rate and the right sides the quality of concentration. However U and V do not represent concentrations and τ does not represent time. The two sides of the equations can be reconciled because these symbols represent only values *associated* with concentrations or time.

Two alternative stable terminal states, Z_1 and Z_2, are possible, representing either a high final value of U and a low one of V (Z_1) or the opposite (Z_2). It is as if the state-space were divided into two separate areas by an imaginary line, which is why the latter is called "separatrix". Theoretically, for all initial states above the separatrix, it is impossible to reach the state Z_2: the trajectories necessarily transfer them to the state Z_1. The opposite applies to all states below the separatrix, which cannot reach the state Z_1. In the immediate vicinity of the separatrix some of the initially closely neighboring trajectories diverge strongly from each other; this is typical of bistable dynamics. If the initial state of the system is located very close to the separatrix, quantum physical effects and thermal fluctuations are able to spontaneously transfer the system's state to the other area of the state-space. Only under theoretical (i.e., purely mathematical) conditions does the separatrix forbid the shift of the system's development from one area of the state-space to another. Under actual conditions – meaning conditions under which random fluctuations exist – such transitions may occur in the immediate vicinity of the separatrix.

The second example refers to the modeling of some interdependent biochemical reactions of the cell cycle. Paning et al. published a model of a simple network of three interconnected reactions from the cell-cycle of a frog's egg (2007, 498). It is worth noting that the computation of the self-organization of this nonlinear system with only three dynamic quantities requires 13 parameters – that is, 13 quantities which take no part in the dynamics of the modeled self-organization. For modeling the cell cycle of yeast the same authors use 36 coupled differential equations – i.e., 36 variables – on which they impose 143(!) parameters (ibid. 499).

Finally, in 2012 a group of biophysicists and bioengineers of the Stanford University and the Craig Venter Institute published a whole-cell computational model of the bacterium *Mycoplasma genitalium* that "includes all of its molecular components and their interactions" (Karr et al. 2012, 389). The model "includes more than 1900 experimentally observed parameters" (ibid. 391). Most of them "were implemented as originally reported" in "over 900 publications" and "several other parameters were carefully reconciled" by the model makers themselves (ibid.).

1.4. Why dynamic systems theory cannot describe organismic dynamics

It is typical of all mathematical accounts of self-organized behavior with which I am familiar – whether in physics, chemistry, or biology – to *essentially depend on a high number of externally set parameters, some of which symbolize gradients*. The sharp division between variables and parameters is not problematic within physics. In real inorganic systems those quantities (which in the models are represented by parameters) cannot be influenced by the system's dynamics itself. In the solar system, for example, neither the mass of the sun nor the gravitation constant – both of which in the models are represented by parameters – are affected by the positions and the movements of the planets. It is obvious, however, that this logic does not apply to organic processes. *In biological systems global factors that influence system dynamics are often internally generated.* For example, although cell volume is an important factor affecting cellular chemistry, the variation of cell volume during the cell cycle is internally regulated. Hence, it is evident that even the most primitive organisms exhibit dynamical characteristic that cannot be modeled by the theory of self-organization. In strong contrast to formal models, the quantities in real networks within organisms are highly dependent on the network's own inner dynamics. In order to preserve themselves in the face of deteriorating conditions, organisms trigger multiple adaptive changes in the factors affecting their internal dynamics (Falkner and Falkner (in this book); Plaetzer et al. 2005). In modeling them as dynamic systems, these changes ought to be described as *internally* controlled changes of parameters if the model makers claim to have created a realistic model of the organism's internal causality. *A model which realistically mirrors the organism's autonomy must be able to calculate at least a significant number of its parameters.* It must be able to independently calculate and adjust also (but not only) those parameters which describe the organism's exchange of energy and materials with its environment. All organisms are constantly doing this. It is a core defining feature of organism!

Therefore, with regard to the sufficiency of dynamic systems theory for biology, the crucial question is whether the sharp distinction between dynamic and static quantities – variables and parameters – that characterizes current formalisms of self-organization actually misrepresents organismic dynamics.

1.4.1. Theoretical evidence

As outlined above, in mathematical models and computer simulations of dynamical systems, parameters are quantities which constrain the development of the systems' dynamics (see section 1.1). Simulation of a fairly autonomous dynamics that could be a model of real organisms would therefore demand that internal dynamics be able to modify a significant fraction of the constraints (parameters) affecting this dynamics. This would be a self-constraining or self-constrained dynamics.

Thus, the minimum requirement that the modeling of real – i.e., biological – self-organization would have to fulfill is that the model calculates how the organism regulates the energetic-material gradients that it depends upon. In mathematical models of self-organization these gradients are represented by externally set parameters. But within current physics this possibility is *excluded*, primarily because all inorganic dissipative dynamic systems have an inherent *entropic tendency* (see section 1.2). They are organized by the gradients imposed on them and so their organization cannot regulate those gradients.

There is another principal reason why a dynamical system's model ignores the possibility for parameters to be calculated in the same way as variables: *Differential equations require a sharp distinction between variables and parameters.* Therefore, within contemporary formalism, it will probably never be possible to dynamize parameters. For mathematical reasons, no formal system is able to compute these quantities within the system.

Often this criticism is rejected on the grounds that many processes in real organisms take place under constant conditions as well. This is, of course, the case, but the constant-holding of such conditions is something that cannot be taken for granted; on the contrary, it is an *achievement of the organism itself*: its *overall dynamics* holds certain quantities at least nearly constant. In the formalism of dynamic systems theory this means that "the overall dynamics of the system repeatedly generates nearly the same value for particular variables".

To put it in a nutshell: For physical and mathematical reasons within theories of dynamic systems or self-organization there is an insurmountable strict distinction between dynamic state variables and externally set parameters – i.e., between constrained and constraining quantities.

It is not only the present author who doubts the possibility of a mathematical formalism which describes a dynamics capable of constraining its own constraints. A prominent founder of systems biology, Stuart Kauffman, writing on the essential limitations of modern scientific formal descriptions of living organisms, says:

> "Consider a cylinder with a piston inside and a compressed working gas between the piston and the cylinder head. The gas can expand, doing work on the piston, pushing it down the cylinder. What are the *constraints*? Evidently the cylinder, the piston, and the location of the piston inside the cylinder, with the gas trapped between the two. But where did those constraints come from? Well, it took work to make the cylinder, work to make the piston, and work to put the gas into the cylinder and the piston in afterward. [...] It appears to take work to make constraints and constraints to make work! [...] [T]he released energy that does work can be used to construct more constraints on the release of energy, which constitutes more work, which in turn constructs more constraints. *Note that these notions are not in the physics or chemistry we have been taught*. One begins to have the sneaking hunch that all this constraint construction on the release of energy – which, as work, can construct more constraints on the release of energy – has something profound to do with an adequate theory of the organization of processes. *We have as yet not even the outlines of such a theory* [...] Nor is the point I am making merely rhetorical. A dividing cell does precisely what I just said. [...] This organization of process is carried out by any dividing cell, yet it is stunning that we have no language – at least, no mathematical language of which I am aware – able to describe the closure of process that propagates as a cell makes two, makes four, makes a colony and, ultimately, a biosphere. [...] [T]he way Newton, Einstein, Bohr, and Boltzmann taught us to do science is limited". (Kauffman 2002, 132-136; italics by S.K.)

Kauffman complains of nothing less than the inability of standard physics and chemistry to describe a self-constraining process – one whose dynamics regulates many of the constraints it (the dynamics) requires, i.e., many parameters (including energetic gradients). Describing such a high degree of self-referential causality is a task for which current mathematical languages are ill-equipped. Systems governed only by efficient causes – that is, systems without any mental capacities – but able to influence most of their parameters would exhibit *an enormous number of causally indefinite states* in their state-spaces. Such systems would be unstable to a much higher degree even than dynamic systems whose state-spaces have areas where closely adjacent trajectories tend to diverge strongly (see fig. 5). Instability is a phenomenon often encountered in the theory of dynamic systems and in systems biology – even if certain parameters are firmly set. The bistability diagrammed in fig. 5 is a kind of instability which occurs even though the parameters are set to fixed values by the model makers.

Because of the *entropic tendency* inherent in an inorganic (and non-mental) dissipative dynamic system the more the system is able to attain influence over its parameters – especially those representing energetic-material openness, i.e., the existence of gradients – the more it will become disorganized. Such a system would tend to continually increase the system's *internal* entropy, i.e., to a permanent increase of the number of its possible states (fig. 6). The entropic tendency of a dynamic system which is governed entirely by non-mental (blind) causes can only be limited if its parameters remain *externally* controlled.

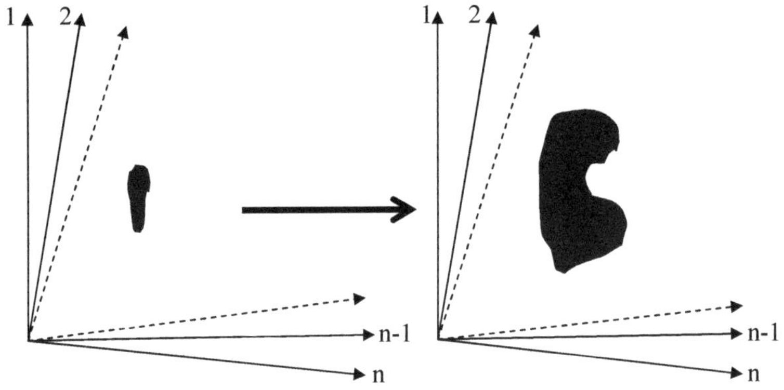

Fig. 6: The increase of entropy within a system depicted as the increase in the number of possible states.

Figure 7 shows another way of describing the increase of entropy – caused by the lost of constraints – as the increase of the number of possible trajectories. If figure 7 represents a model of a real organism's dynamics, *only a very limited number of these possible trajectories would be biologically viable*; in other words, *very few would represent states of being alive*. This is the case since though biological structures are physicochemical structures they constitute only a vanishingly slight number of the latter. In fig. 7 the long curve represents a thin bundle of biologically viable trajectories, while the dotted lines stand for developmental trajectories which are possible in terms of physics and chemistry, but fatal from a biological point of view. They involve the derailment into areas of increasing entropy or uncertainty.

This leads necessarily to the following aporia: How does an organism succeed in avoiding *derailments* into areas of disorder if it often faces possibilities equally valid from the point of view of physics and *lacks any capacities* allowing a biologically adequate choice between these possibilities?

In the past the answer might have been "the genes" or "genetic information". But both these notions are much less clear today than they were some decades ago. Now we understand that genes are massively co-determined by the organism's dynamics.

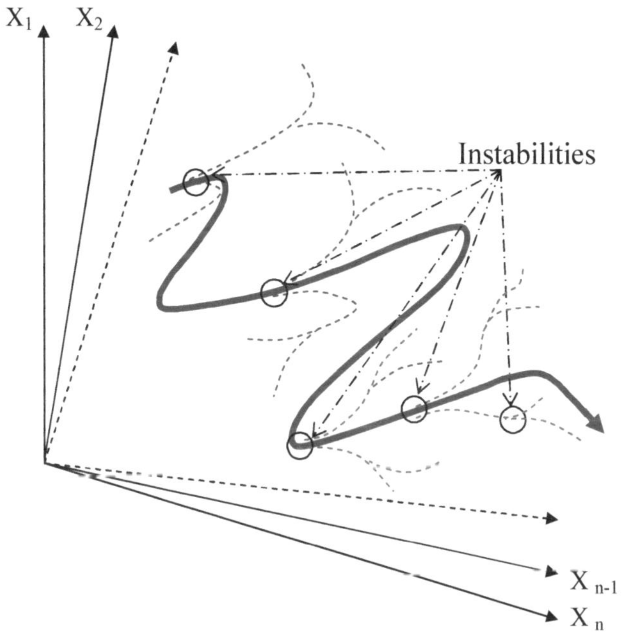

Fig. 7: Permanently occurring instabilities in the development of a hypothetical dissipative dynamic system which influences the value of its parameters entirely by non-mental (blind) forces. If understood as displaying a real organism, the variables X_1 to X_n represent important dynamical quantities (like concentrations of proteins and signal substances etc.), the coherent variation of which is characteristic of an organism. The long curve and the dotted lines do not represent single trajectories but rather bundles of these.

1.4.2. Experimental evidence

There is also experimental evidence demonstrating that the causal order of organisms remains beyond the reach of a standard physical model of self-organization. Biologists have known for a long time that the exchange of energy and material between real organisms and their environment can*not* be understood in terms of increasing entropy production. This follows because a high rate of entropy production means a high rate of the degradation or waste of energy. The experimental and theoretical research of Gernot and Renate Falkner clearly shows that the metabolic exchange between cyanobacteria and their environment only exhibits a high rate of entropy production if the physiological adaptation of the bacteria to their environment has been disturbed and is merely being readjusted (Falkner and Falkner, this volume). The act of physiological re-adaptation, as they describe it, effects a *decrease* of entropy production and not its increase. This is the opposite of what is to be expected from a physico-chemical theory of self-organization. Only an organism which is no longer in a state of optimal adaptation seems to function in a way consistent with the physical conception of self-organization. When out of balance with its environment this biological system is in a state of low entropy but produces entropy at a high rate. But as soon as re-adaptation is reached, the organism is situated at a state of low entropy and low entropy production as well. This fact contradicts the predictions of standard dynamical systems models. So, while the act of physiological adaptation begins as if the organism were following the laws of inorganic self-organization, in the latter stages it proceeds in a manner that can only be explained in biological terms. It is plausible that all organisms tend to develop toward a minimization of entropy production. Physiological re-adaptation is a biological act that requires the *internally conditioned* and coordinated variation of many dynamic quantities which, in the models of systems biologists, are described as fixed parameters.

There is also experimental evidence showing that a process of minimization of entropy production takes place during embryonic development. The results of numerous measurements in several studies have led to the conclusion that the entropy production "indeed decreases at separate stages of the ontogenesis (if early stages of development are excluded)" (Martyushev and Seleznev 2006, 40).

1.5. Summary and conclusion of section 1 – the essence of the organism

Every organism must act anti-entropically – i.e., maintain or reduce its low level of entropy – *without* requiring most of the conditions of its self-organization to be externally set. Both theoretical considerations and experimental evidence make clear that *the theory of self-organization of physics is too weak to account for real biological self-organization* (Koutroufinis 1996). From this point of view the essential feature of an organism is that its dynamics constrains itself whereas the dynamics of inorganic self-organized systems do not. Organismic dynamics does not require that a large number of constraints are externally given as they are in inorganic systems. In other words, organisms are able to generate most of their critical constraints autonomously.

To fail to see the limitations of the theory of dynamic systems – and thereby consider the models of systems biology as appropriate descriptions of an organism's internal causality – is a clear case of Whitehead's "fallacy of misplaced concreteness", since it confuses something abstract (model) with something concrete (organism).

Nevertheless the theory of dynamic systems allows us to make a crucial assumption about organismic dynamics. *A system which has the complexity of an organism and is ruled only by efficient causation would face a huge number of equally valid possibilities in its development.* Of course, only very few of these are biologically viable.

This result invites us to go beyond the ontological limitations of current scientific teleology. Since it is not capable of accounting for anti-entropic behavior in real organisms, we should consider conceptions of teleology which go beyond the confines of physics. We should consider the relevance of causal factors which are not efficient causes in organisms. We need a form of teleological causality which does not violate the efficient causality of physics and chemistry, but coexists with it, for it is obvious that laws of physics and chemistry remain valid in organisms.

2. Mentalistic teleology and process philosophy

Scientific teleology is founded on metaphysical presuppositions entirely different from those of mentalistic teleology. The fact that many dynamical systems of interest to science and technology reach a terminal state in an abstract state-space (see fig. 1 and 5) at a later time can be exhaustively explained by the material structures of the systems that were *physically present* at an earlier time. The same is the case for the convergence of the operations of a computer towards a terminal state.

The idea of a mental teleological factor, on the other hand, is rooted in two main presuppositions:

- Firstly, that there are causal factors oriented towards something *physically absent.*[14]
- Secondly, that the operations of such factors cannot in principle be exhaustively explained by referring to the material structure of something physically present.

Mentalistic-teleological factors do not influence a system's development from the future by "pulling" it to a particular goal, as has often been (wrongly) claimed. They act in the present: they *anticipate*. Anticipation is a mental operation of the anticipating subject occurring in the present – it has nothing to do with time-reversal. *Mental teleology is there if a subject in its present is anticipating for its own future something physically absent in the present.* My understanding of this idea is not at all metaphorical. Anticipation is *experienced* by the anticipating agent; it has an *inner* side, a quale, as do all our mental acts. So far science has not succeeded in reducing the qualia of mental acts to physically or spatio-temporally present facts.

However, mentalistic teleology should *not* be taken to ascribe consciousness to cells, unicellular organisms, plants and lower animals. Muraca's clear distinction between intentionality (Zwecksetzung, ZS), which is the conscious setting of an aim, and agency (Zwecktätigkeit, ZT), which does not necessarily imply consciousness, helps to avoid this common misconception (Muraca, this book: section 2.3). Mental activities refer *to* something; for example, the perceptions of heat. But only in very rare cases can they be likened to human intentionality and awareness. In almost

[14] I owe the concept of *absence* to Terrence Deacon, professor for biological anthropology at the University of California, Berkeley. He has introduced the term to current discussions about teleology, function and information (2007). However, in contrast to my process philosophical understanding of organismic teleology Deacon rejects the relevance of any kind of protomentalistic factors in organisms (2012, 77-79).

all cases mentalistic teleology takes place entirely within the realm of *unconscious* sentience. By "sentience" I mean a very specific form of making distinctions on the basis of qualia. It is an apprehension of reality rich in qualitative contrasts which stresses some of its aspects while ignoring others. The qualia of very simple living beings like unicellular organisms appear to be undifferentiated experiences of sympathy and antipathy.

In my opinion this conception of teleology is tied to at least two conditions:

- Firstly, in contrast to the convictions held within *classical* physics, the (spatiotemporally present) physical side of the organism attains causal relevance through the mediation of something else. This "something" is correlated to the physical side of the organism, but not determined by it. The non-physical side of the organism can be thought of as the "organismic subject".
- Secondly, and closely connected to this is, that which is physically present allows the physically absent organismic subject to act in different ways. This means that the physical state of the organism at a specific moment permits the realization of different physical states at the next moment (see fig. 7).

Both these conditions must be met, because talking of "subjectivity" and "anticipation" makes no sense if the future is already fixed in the present.

I consider Whiteheadian process-metaphysics to be conducive to this conception of mentalistic teleology.

2.1. A brief excursus on the metaphysics of Alfred N. Whitehead

The ontological basis of Whitehead's natural philosophy has been outlined in the introduction to this volume. Here I will summarize his metaphysics by discussing six points important to the topic of ontogenesis:

1) "Process" is the central concept of Whiteheadian ontology. Whitehead does not call changes or movements "processes", but only individual or indivisible events. Another word for what Whitehead calls process is "actual occasion" or "actual entity".

2) All actual occasions have a double nature: a mental-physical or subjective-objective bipolarity. All processes are acts of determination of their own nature or essence. The subjective or inner side of processes consists in their striving for self-determination. Actual occasions manifest the result of their self-determination in space and time in order to express their subjective or inner side. As space-time-data they may become objects of other actual entities which are still in the act of becoming, i.e., still in the subjective phase of their existence. So, completed actual occasions can be integrated in later (still becoming) processes or subjects.
3) It is essential to the Whiteheadian theory of causality that every physical or spatiotemporal fact, even the simplest quantum event in an atom, can only have an effect after it is taken in by a still non-completed actual entity. Whitehead's term for this inclusion or absorption of an object in the self-determination of a new subject is "physical prehension". This concept is related to the ordinary concept of apprehension, but means the most elementary form of perception and experience in nature and is in very few cases accompanied by consciousness. Not only the fact of this absorption (through prehension) is determined by the absorbing subject itself, but also *how* the absorbed object will be integrated into the absorbing subject. Whitehead emphasizes the creativity of all actual occasions.
4) The many already constituted actual occasions which are prehended by a becoming actual occasion as its objects and absorbed into its own self-creation[15] provide the becoming process with a spectrum of different possibilities for its actualization: The physical constitution of the immediate past of a becoming actual entity (i.e., the material constitution of its prehended objects) determine certain possibilities for the self-creation of the new process. The objects do not offer more than "real potentialities", as Whitehead calls them, without being able to determine which of them will be actualized by the new process.
5) The subjective side of the becoming process creates the essence or, as Whitehead calls it, the "real internal constitution" of the process through an act of decision between the real potentialities: "[...] 'decision' is the additional meaning imported by the word 'actual' into the phrase 'actual entity'. *'Actuality' is the decision amid 'potentiality'. The real internal consti-*

[15] See introduction of this volume: section 3.2.

tution of an actual entity progressively constitutes a decision [...]" (Whitehead 1979, 43, italics by S.K.). The manifestation of the completed (fully determined) actual occasion as a spatiotemporal datum is its objectification – the expression of its act of decision between real potentialities.
6) The process reaches its terminal state only after the decision has eliminated all uncertainty concerning the actualization of real potentials. Only then does the new actual occasion appear as a spatiotemporal datum. Accordingly, the end of a Whiteheadian process is a "jump" in the space-time of physical reality. Its physical manifestation is extremely short-lived; it is more like a flash of lightning. The micro-physical and micro-chronic events of quantum-physics (e.g., electronic and photonic events), the simplest subjects according to Whitehead, disappear after a 10^{-x}-second-lasting presence in space (where x is considerably bigger than 3 and at most equal to 43). The most complex and long-lasting processes are the acts of human consciousness, which can last up to a maximum of a few seconds and manifest themselves as macroscopic patterns of neuronal activation in the brain.

2.2. Decisions for life: living occasions

Whitehead considers the macroscopic enduring things of our day-to-day experience to be *societies* of actual occasions. Although almost all actual occasions have a *micro-chronic* presence in ordinary three-dimensional space, their extremely fast sequence allows societies to appear as *macro-chronic* (i.e., persisting) things. Most societies are very simply organized, so that the actions of their microscopic processes do not support each other mutually. No *macroscopic* coherent activity emerges in these societies. But some societies are organized in such a way that their actual occasions do support each other. They are the most dynamic and causally prevailing parts of the bodies of living beings; Whitehead calls them "living societies". The mutual support of microscopic processes has the effect that the whole organism acts like a single macroscopic process: it autonomously takes in matter and energy and determines their causal relevance. In sharp contrast to the self-organized systems of physics, living societies impose on themselves material and energetic gradients, for example by searching for and taking on matter and energy in the form of food by themselves.

For Whitehead the body of an organism is just the external, objective side of a society of highly complex mental-physical processes appearing in space-time. In his metaphysics, organisms are considered as societies of processes called “living occasions” (1979, 104).[16] Living occasions do not ontologically differ from other actual entities. Their peculiarity consists only in that they introduce something to the life of an organism that has not been realized ever before in its past. It is characteristic for living occasions to not conform to the past history which they inherit. They introduce something new to the history of a living being which cannot be totally explained as result of its past.

As mentioned above, the organismic subject is the factor which lends causal relevance to the organism’s physical side. It prehends the organism’s most recent state and creates the succeeding state by its own manifestation or objectification. Thus process philosophy does not regard the physically or spatiotemporally manifest part of the organism as a gapless chain of physical states in which each state automatically determines its successor state.

Let us return to the aporia concerning the avoidance of derailments into areas of disorder (fig. 7). Like all actual entities every living occasion has a mental and a physical side. Its mental side recognizes the possibilities of the very near future and makes a choice which is biologically viable. Then its physical side manifests this choice as a material datum in space-time and thus determines the state of the organism. The “regions” of the state-space where living occasions can act are the unstable states in the development of the organism (fig. 7). The concept of living occasions provides the following solution: *A single living occasion can provide an organism which is at the very beginning of the divergence of neighboring trajectories in an unstable area of its state-space with a biologically viable direction.*

By means of its manifestation as a spatiotemporal datum with a certain material constitution, a single living occasion is able to keep an organism whose development passes through indetermined areas of the state-space away from states that are biologically not viable (fig. 8). The physical constitution of the immediate past of a becoming living occasion – i.e. the material constitution of that part of the organism which has been prehended by this

[16] See introduction of this volume: section 3.5.

process – provides the new living occasion with various possibilities for the creation of its own (newly synthesized) material constitution. Thus, the physical state of an organism at a certain point in time and the laws of nature allow a range of possible developments to occur in the immediate future of the organism. The living occasions that take place at this point in time will actualize only one of these real potentialities, which they will do by their own manifestations as the new material facts of the organism. The peculiarity of living occasions consists in their deciding upon developmental paths that are *biologically viable* and not just physically and chemically *possible*. Such important decisions cannot be made by ordinary actual occasions.

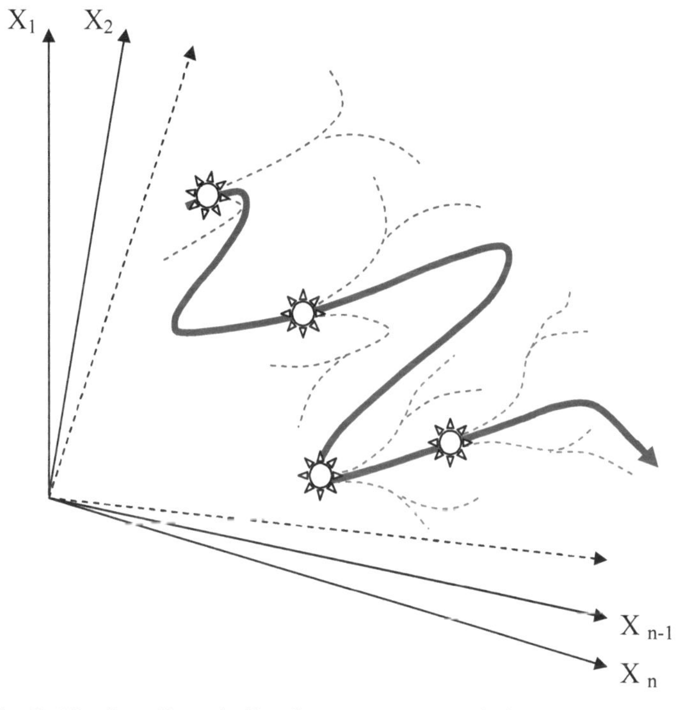

Fig. 8: The "suns" symbolize *living occasions*. Their decisions actualize very few physico-chemical real potentialities which are biologically viable: they prevent the organism from derailing to the lethal developments represented by the dotted lines, thus keeping it on a biologically viable track.

Entropy is uncertainty.[17] Since living occasions reduce uncertainty through the exclusion of almost all of the possible future states of the organism they are essentially anti-entropic agents. They protect embryogenesis from teratogenesis (i.e., monstrosity) and control the regeneration of the mature organism in the event of injury or disease.

Whiteheadian natural philosophy has inspired physicists like Roger Penrose (1995, 1994, 1989) and Henry Stapp (this volume, 2004) and the physician Stuart Hameroff (2007, 2003, 1996). Their writings are of particular interest for the discussion of living occasions in terms of current quantum physics. They, along with other scientists, argue for the existence of actual occasions of *mesoscopic* size which play a decisive role in biological events (see also Gunter, this volume). Thanks to their size these processes would easily be capable of moving the state of an organism onto a certain trajectory which is both possible from the point of view of physics and biologically viable, thereby preventing the derailment of the organism to areas of high entropy. Such quantum events would not be blind quantum fluctuations – as they are actual occasions – but directed *and* mental acts; they would be mental-teleological processes. In this connection, it is noteworthy that a lot of progress has been made recently in the explanation of photosynthesis by applying the idea of quantum events that remain coherent over mesoscopic distances and non-microchronic timescales.[18]

Quantum theory allows us to make a very important clarification at this point: The last figure and figure 7 should not be interpreted in terms of classical or statistical mechanics. Because of Heisenberg's uncertainty principle, the development of the organism should not be considered as occupying a certain point in the state-space at a certain point in time as these figures suggest. At a point in time it occupies a very small volume of the state-space, but because of Heisenberg's uncertainty principle this volume cannot be infinitesimally small. The organism's development can be portrayed by a sequence of such volumes. Thus Whitehead's rejection of the scientific abstraction of "simple location" (1953, 61f.) may also be applied to the depiction of the organism's development in state-space, although it

[17] See section 1.2 and fig. 3 and 6.
[18] See Collini et al. 2010, Engel et al. 2007, Lee et al. 2007.

is important to point out that his criticism does not follow from the uncertainty principle but from his own idea of prehension (ibid. 86f.).

Figure 8 often leads to another misunderstanding which should be guarded against: the figure does not imply that each living occasion has a "map" on which all possible developments of the organism's dynamics, like the streets of a city, are drawn in advance. Only *we* as outside observers are able to draw a "map" containing all possible trajectories, even if only in principle[19] – the biologically viable ones and all the others. Protomental processes like living occasions cannot have such a "map" at their disposal as they are not endowed with consciousness. Thinking of a *decision* as the willful pursuing of a long-term plan fails to recognize that the real meaning of this word in English and other European languages refers to the rejection or elimination of options. Of course, it is possible to reject an entire plan consisting of a tree-like web of succeeding decisions. But this would require a high developed conscious being which has mastered complex logical operations with abstract entities, giving it a large time horizon. The simpler an organism, the more "short-sighted" are its plans. It would be incorrect to say that the living occasions which prehend those parts of an organism just beginning to derail into states of increasing disorder realize this danger by comparing the actual state with the ideal state and measuring the deviation. They do not analyze the material constitution of which they prehend; rather, they *experience* it. Living occasions do not correct the development of the organism like the captain of a ship, who calculates the course anew by using an oceanographic chart and navigational instruments. Rather, they are led by something which may be described as remembering the experience of the state of being healthy. The organism corrects its embryogenesis by means of its inner perspective or self-experience, its *inwardness*, which it possesses even as an embryo. Every multicellular organism has an embryogenetic and an immunological memory. The latter is mainly individual as it results from experiences encountered during ontogeny which only occurs once; the former is supra-individual as embryogenesis is rooted in the species.

---

[19] I use "in principle" to mean "if we had a perfect science (physics, biochemistry, biophysics etc.), knew all the details about the physical constitution of the organism and had unlimited computer-power".

It is not possible to say more about this topic here for this would require focusing on the intricate query of the rootedness of embryogenetic memory in the species to which the organism in question belongs. Also, other essential questions cannot be addressed here: What coordinates the contemporaneous, and causally independent, living occasions which prehend different parts of the organism so that their decisions generate a coherent result which looks like the decision of a single subject that could be called the "organismic subject"? Would it make sense for each living being to hypothesize a single overall organismic subject with its own inwardness which would be the source and hence the coordinator of the short-lived living occasions? Could this long-lived organismic subject be conceived as an actual entity?[20]

2.3. Processual teleology

Whitehead regards every becoming of an actual entity as a "teleological self-creation" (1967, 195). He does not, however, reduce teleology to the concept of function in the way that neo-Darwinists and most of the contemporary philosophers of biology who accept the concept of teleology do.[21] The essence of a living occasion consists in a powerful decision; this decision is powerful because it is, to a significant extent, anti-entropically effective. Such decisions are protomental acts which have at least an elementary capacity for experience. The subjective side or mental pole of the living occasion experiences its own tendency towards one possible development and its rejection of all the others with the phenomenal qualities (qualia) of sympathy and aversion respectively.

Whiteheadian ontology is based on the original and genuine understanding of "telos" which avoids the confines of functionalism. All kinds of end-states in the development, readaptation, behavior, regeneration etc. of organisms can only then be reached if they are striven for as something

[20] I have dealt with these questions in my book *Organismus als Prozess* (forthcoming).
[21] On the reduction of teleology to function in neo-Darwinism see Mayr 1991, 75, 61; Brandon 1990, 188; Ariew 2007, 179. The most recent kind of biological neo-teleologism is theoretically founded on dynamic systems theory, that is on theories of self-organization and complexity (Koutroufinis 2013, 314, 318-327), Christensen 1996).

positively experienced. Aristotle's teleology goes beyond functionalism as well. His world view simply *forbids* considering a natural process controlled by blind, i.e., non-mental forces, as being able to achieve the kind of ordered result attained by an appropriately formed organic structure that serves the purpose of staying alive, like an organism or an organ, rather than degenerating into chaotic malformation (*Physics* II, 198 b33-199 a1). Aristotle would never assume that non-mental processes would be able to produce something as ordered as a single cell.

Another common position of Aristotle and Whitehead is that teleology is a mentalistic notion which does not presuppose the idea of conscious agency. It is crucial to Aristotelian metaphysics that, in nature, mental agents are only rarely conscious of their acting.[22] Conscious action is only a seldom occurring special case of mental activity.

Despite important similarities to Aristotelian teleology, Whiteheadian natural philosophy and its understanding of *process* introduces a new kind of teleology which is unique in the history of Western metaphysics. It is no accident that one would search in vain for the concept of "entelechy", a term so central in Aristotle's theory of teleology, in Whitehead's works. Whitehead supports a moderate teleology or final-causality conception. His most basic hypothesis is that all elementary processes or actual entities are acts of experience striving towards the ultimate determination of their own essence. This means that an actual entity *does not strive to achieve a predetermined end-state*, as teleological development is often interpreted, but first and foremost to find out what would be an appropriate end-state for itself: in other words, to define the aim of its own self-creation. Central to this is the idea that the crystallization or gradual development of the aim or telos towards which an actual entity strives belongs to its essence: "Process is the growth and attainment of a final end" (1979, 150). In this sense, one may say that Whitehead replaced the in some respects static substance-philosophical teleology of antiquity and the middle ages with a *processual teleology*. His metaphysics offers to biophilosophy a modern basis for a re-

[22] "It is absurd to suppose that purpose is not present because we do not observe the agent deliberating. Art does not deliberate. If the ship-building art were in the wood, it would produce the same results by nature. If, therefore, purpose is present in art, it is present also in nature" (*Physics* II, 199 b26-30).

conception of genuine Aristotelian insights concerning telos and organism. The combination of processuality and mental-physical bipolarity allows a transformation of the Aristotelian concept of telos – as an insoluble connection of terminal state and purpose – beyond both the old metaphysical concept of substance and the neo-Darwinistic reduction of purpose to function.

Whitehead's processual teleology does not suffer from the serious shortcomings of substance-ontology and different versions of vitalism. In particular, by virtue of the mental-physical bipolarity of actual occasions, it is free from the substance-dualism which haunts many approaches, for example psycho- and neo-vitalism.

Another serious shortcoming of both substance-dualism and psycho-vitalism is that they implicitly violate the law of the conservation of energy by introducing nonphysical forces that act upon matter. In contrast, because of abstaining from additional nonphysical entities Whiteheadian ontology is able to abide by the conservation of energy law: The Whiteheadian cosmos is a vibrating one since the most elementary entities of actuality, the actual occasions, are not persisting substances that are permanently present in space-time, but rather flashes of spatiotemporal existence – vibrations of being. Instead of thinking of the "soul" and the "matter" of the organism as two persisting entities or substances which interact, Whitehead considers living beings as special societies, the members of which permanently originate and perish. The material side of the processes of the living being is actualized over and over again with a very high frequency. The living occasions and all the other actual occasions taking place within organisms manifest themselves in space-time as quanta of energy; thus *they are the energy of the organism*. With their mental side they can make two important decisions: firstly, how they will distribute the overall energy which comprises their physical side in the body of the organism and, secondly, with what material constitution, that is to say with what molecular structure it will be manifested there. Thus they do not "push" the constitution of the organism's body into a certain state (into a certain location in the state-space) by any non-material "forces"; this would violate the law of the conservation of energy. Whiteheadian processes do not act on the material world "externally", but rather "internally" through their manifestation *as* matter.

3. Conclusion

Viewing organisms as merely dynamic systems in terms of physics and chemistry cannot explain their real autonomy. Organisms viewed in this limited way would succumb to the serious problem of instability: The organism's dynamics would permanently enter phases in which the trajectories of its development would diverge strongly; furthermore, most of them would derail the organism's development into fatal disorganization. But at the same time the problem of diverging trajectories offers a fruitful application for Whiteheadian teleology. It indicates that the future development of the physical side of the organism is causally open to a certain degree so that a particular kind of Whiteheadian processes, namely living occasions, have possibilities of choice. Living occasions can provide an organism which is at the very beginning of the divergence of neighboring possible developments with a biologically viable direction by means of their manifestation as spatiotemporal data with a certain material constitution. The essence of each living occasion is a decision which has an anti-entropic effect, since it strives to prevent the organism from increasing its entropy.

Bio-systemism and Whiteheadian process-philosophy can be merged into a higher synthesis whose core idea can be formulated as follows: Biosystemic thinking can in principle[23] describe the *possible* developments of an organism, while the organism's actual development, consisting in the decision between real possibilities, is accessible to Whiteheadian process philosophy.

References

Arew, A. (2007). "Teleology". In: Hull, D.; Ruse, M. (eds.). *The Cambridge Companion to the Philosophy of Biology*. Cambridge, New York: Cambridge University Press, pp. 160-181.

Aristotle. *Physics*. The Internet Classics Archive http://classics.mit.edu/Aristotle/physics.2.ii.html

Brandon, R. (1990). *Adaptation and Environment*. Princeton (NJ): Princeton Univ. Press, 1990.

[23] For the meaning of "in principle" in this context see footnote 19.

Christensen, W. (1996). "A Complex Systems Theory of Teleology". In: *Biology and Philosophy* 11, pp. 301-320.

Collini, E.; Wong, C.; Wilk, K. et al. (2010). "Coherently wired light-harvesting in photosynthetic marine algae at ambient temperature". In: *Nature* 463, pp. 644-647.

Deacon, T. (2012). *Incomplete Nature*. New York: W. W. Norton & Company.

——(2007). "Shannon-Botzmann-Darwin: Redefining information. Part 1". In: *Cognitive Semiotics* 1, pp. 123-148.

Ebeling, W. (1976). *Strukturbildung bei irreversiblen Prozessen*. Leipzig: Teubner (in German).

Ebeling, W.; Engel, A.; Feistel, R. (1990). *Physik der Evolutionsprozesse*. Berlin: Akademie-Verlag (in German).

Ebeling, W.; Sokolov, I. (2005). *Statistical Thermodynamics and Stochastic Theory of Nonequilibrium Systems*. New Jersey, London, Singapore etc.: World Scientific Publishing.

Engel, G.; Calhoun, T.; Read, E. (2007). "Evidence for wavelike energy transfer through quantum coherence in photosynthetic systems". In: *Nature* 446, pp. 782-786.

Falkner, G.; Falkner, R. (this volume).

Gardner, T.; Cantor, C.; Collins, J. (2000). "Construction of a genetic toggle switch in Escherichia coli". In: *Nature* 403(6767), pp. 339-342.

Gunter, P. (this volume).

Hameroff, St. (2007). "Orchestrated Reduction of Quantum Coherence in Brain Microtubules A Model for Consciousness". In: *NeuroQuantology* 5 (1), pp. 1-8.

——(2003). "Consciousness, Whitehead and Quantum Computation in the Brain: Panprotopsychism Meets the Physics of Fundamental Space-Time Geometry". In: Riffert, F.; Weber, M. (eds.). *Searching for New Contrasts*. Frankfurt a.M.: Peter Lang, pp. 61-86.

Hameroff, St.; Penrose, R. (1996). "Conscious Events as Orchestrated Space-Time Selections". In: *Journal of Consciousness Studies* 3, No. 1, pp. 36-53.

Hull, D. (1974). *Philosophy of Biological Science*. Englewood Cliffs, New Jersey: Prentice Hall.

Karr, J.; Sanghvi, J. et al. (2012). "A Whole-Cell Computational Model Predicts Phenotype from Genotype". In: *Cell* 150 (2), pp. 389-401.

Kauffman, S. (2002). "What is Life?" In: Brockman, J. (ed.). *The Next Fifty Years*. New York: Vintage Books, pp. 126-141.

Koutroufinis, S. (2013). "Teleodynamics: A Neo-Naturalistic Conception of Organismic Teleology". In: Henning, B.; Scarfe, A. (eds.). *Beyond Mechanism: Putting Life Back Into Biology*. Plymouth UK: Lexington books, pp. 309-342.

——(forthcoming). *Organismus als Prozess.* (in German)

—— (2012). "Organismus-Verständnis zwischen Zweckmäßigkeit und Entropie – eine historisch-theoretische Betrachtung". In: Gadebusch-Bondio, M; Siebenpfeiffer, H.; Fischer, T. (eds.): *Konzepte des Humanen*, Freiburg, München: Alber, pp. 191-206 (in German).

—— (1996). *Selbstorganisation ohne Selbst*. Berlin: Pharus-Verlag (in German).

Lee, H.; Cheng, Y.; Fleming, G. (2007). "Coherence Dynamics in Photosynthesis: Protein Protection of Excitonic Coherence". In: *Science* 316 (no. 5830), pp. 1462-1465.

Martyushev, L; Seleznev, V. (2006). "Maximum entropy production principle in physics, chemistry and biology". In: *Physics Reports* 426, pp. 1-45.

Mayr, E. (2000). *Das ist Biologie*. Heidelberg, Berlin: Spektrum Akademischer Verlag (in German).

—— (1991). *Eine neue Philosophie der Biologie*. München, Zürich: Piper (in German).

Muraca, B. (this volume).

Nicolis, C; Nicolis, G. (2010). "Stability, complexity and the maximum dissipation conjecture". In: *Quarterly Journal of the Royal Meteorological Society* 136, pp. 1161–1169.

Normile, D. (1999). "Building Working Cells 'in Silico'". In: *Science* 284/1999, pp. 80-81.

Panning, T.; Watson, L.; Shaffer, C.; Tyson, J. (2007). "A Mathematical Programming Formulation for the Budding Yeast Cell Cycle". In: *SIMULATION* 83, pp. 497-513.

Penrose, R. (1989). *The Emperor's New Mind*. New York: Penguin Books.

—— (1994). *Shadows of the Mind*. Oxford, New York: Oxford University Press.

—— (1995). "Why new physics is needed to understand the mind". In: Murphy, M. P.; O'Neill, L. A. J. (eds.). *What is Life? The next fifty years. Speculations on the future of biology*. Cambridge: Cambridge University Press, pp. 115-130.

Pittendrigh, C. (1958). "Adaptation, Natural Selection and Behavior". In: Roe, A.; Simpson, G. (eds.). *Behavior and Evolution*. New Haven: Yale University Press, pp. 390-416.

Plaetzer, K.; Thomas, S.; Falkner R.; Falkner G. (2005). "The microbial experience of environmental phosphate fluctuations. An essay on the possibility of putting intentions into cell biochemistry". In: *Journal of Theoretical Biology* 235, pp. 540-554.

Ross, J.; Corlan, A.; Müller, S. (2012). "Proposed Principles of Maximum Local Entropy Production". In: *The Journal of Physical Chemistry B* 116, pp. 7858-7865.

Sagan, D. (2008). "Evolution, Complexity, and Energy Flow". In: Cobb, J. B. Jr. (ed.). *Back to Darwin. A Richer Account of Evolution*. Grand Rapids (MI), Cambridge (U.K.): Eerdmans Publishing Co., pp. 145-156.

Salthe, S. (2010). " Maximum Power and Maximum Entropy Production: Finalities in Nature". In: *Cosmos and History: The Journal of Natural and Social Philosophy,* 6 (1), pp. 114-121.

Schneider, E.; Kay, J. (1995). "Order from Disorder: the Thermodynamics of Complexity in Biology". In: Murphy, M.; O'Neill, L. (eds.). *What is Life? The Next Fifty Years*. Cambridge: Cambridge University Press, pp. 161-173.

Stapp, H. (2004). "Whitehedian Process and Quantum Theory". In: Eastman, T.; Keeton, H. (eds.). *Physics and Whitehead.* Albany: State University of New York Press.

——(this volume).

Swenson, R. (1997). "Autocatakinetics, Evolution, and the Law of Maximum Entropy Production: A Principled Foundation Towards the Study of Human Ecology". In: *Advances in Human Ecology* 6, pp. 1-47.

Toepfer, G. (2005). "Teleologie". In: Krohs, U.; Toepfer, G. (eds.). *Philosophie der Biologie*. Frankfurt am Main: Suhrkamp, pp. 36-52 (in German).

Tomita, M. (2001). "Whole cell simulation". In: *Trends in Biotechnology* 19(6), pp. 205-210.

Wayt Gibbs, W. (2001). "Simulierte Zellen". In: *Spektrum der Wissenschaft* 11, pp. 54-57 (in German).

Whitehead, A. N. (1979). *Process and Reality*. New York: Free Press.

——(1953). *Science and the Modern World.* Cambridge: University Press.

——(1967). *Adventure of Ideas*. New York: Free Press.

Wolpert, L. (1995). "Development: Is the Egg Computable or Could We Generate an Angel or a Dinosaur?" In: Murphy, M.; O'Neill L. (eds.). *What is Life? The Next Fifty Years*, Cambridge: Cambridge University Press, pp. 57-66.

Process and Action: Whitehead's Ontological Units and Perceptuomotor Control Units

JONATHAN T. DELAFIELD-BUTT

> "Motion in the most general sense, conceived as the mode of existence, the inherent attribute of matter, comprehends all changes and processes occurring in the universe, from mere change of place right to thinking." Friedrich Engels, *Dialectics of Nature*, 1888

Introduction

Movement lies at the heart of matter; all matter is always moving, in many different modes. Matter organised in the form of living cells and animals does something peculiar, however. Its collective movements are *purposeful* and *goal-oriented*, geared for the appropriation of an immediate future and generally for the preservation of its existence. The organisation of the movement of matter lies at the heart of living: from the organised biochemical activity within small compartments of cells, to the movements and activities of cells and cell systems, to the gross body movements of large multicellular organisms like you and I. All of life is the organised movement of matter.

Living organisation enacts purposeful behaviours that sustain the existence of the living thing, a feature that I will show here fundamentally requires prospective control to anticipate the future present. How an organism senses and moves prospectively purposefully within its environment is an essential question for the emerging "systems" sciences, and is the theme for this paper.

I will present two separate streams of thought that are closely analogous. The first is the process metaphysics of Whitehead's "actual occasion" defined in his "philosophy of organism" (Whitehead 1979 [1929]). The second is a perceptuomotor control theory from ecological psychology (Gibson 1966, 1979; Reed 1996) based on the "*tau*" informational variable, "general *tau* theory" (Lee 1998, 2005). General *tau* theory explains the goal-directed processes of animals perceiving and acting with intention. Similarly, Whitehead's actual occasion explains the goal-directed process in a prehending and acting ontological entity. They both explain a process of sensing, integrating, and acting in the world, but where one explains this process occurring through spacetime in a living animal, the other considers the process as a fundamental ontological construct. The juxtaposition of the two helps to inform each theory and so broaden our understanding of some important notions, namely the component elements of the ontological unit and the psychophysical construct of a perception-action cycle, as well as leading us to other notions, such as what "process", "mind", and "time" are.

The fundamental similarities between Whitehead's ontological unit and a unit of action in general *tau* theory are: (1) they are both goal-directed, obtaining data in the "specious" present for a *prospective* purpose; (2) they both integrate and re-integrate new data into the single act to achieve their goal;(3) they are both guided by a "subjective aim", which determines the course of the integration and the act; (4) they are both atomic units individually discreet from each other, but in practice (5) they are interwoven with many others to form "fluid", coherent wholes. Finally, (6) they both involve a mental pole, which is the perceptual and intentional part, together with a physical pole, which is the acting out of the subject into the material of action.

Both Whiteheadian thinking and "*tau*" thinking are also firmly embedded in ecological thinking, the former helping to give rise to ecological thought in the early 20th Century and the latter an important contributor to "ecological psychology", a branch of psychology that considers the importance of an experiencing agent acting in active engagement with an eventful, changing world. Their units of construction and the way these elements are arranged are also remarkably similar, so similar in fact, to be considered almost analogous. I will arrange each in the form of an elementary control system based on Weiner's fundamental cybernetics (Figure 2), a study of the fundamental physical laws of process, to illustrate this clearly.

1. Ontological units: Whitehead's actual occasion

Whitehead's unit is the actual occasion: the "fundamental drop of experience" (Whitehead 1979 [1929], 18 [28]). It is both matter and mind together. It is one unit of matter and mind acting outward into the world and at the same time integrating its sense or perception of the world. The actual occasion is "what there is", the "thing-in-itself". Experience as we know it, according to Whitehead, is a society of these actual occasions, one succeeding the next and inter-relating with others. Each one prehends its world, which is composed of the other actual occasions that are objectifying around it, together with its own subjective form. These senses, or prehensions as Whitehead calls them, are integrated according to the subjective aim of the actual occasion. The subjective aim guides the process of prehending and integrating the prehensions. It continues to do this until the point of satisfaction is reached. Satisfaction is the result of all the data prehended, integrated, and acted out into the world in objectification with others, which is the concreteness of the actual occasion.

The objectification is the "material", it is the acting out of the actual occasion in satisfaction. It is what the other actual occasions can know of this actual occasion. Objectification is the public expression of one actual occasion's satisfaction, while the prehending and integrating constitute the "mental" pole.

I have dissected out this ontological unit from Whitehead's main body of work, which dissociates it from the complexities of his larger theory to better study it and understand it for what it is. I have removed many of the refinements of the elements, such as "eternal objects", differences of "prehension", and the many complexities of "subjective form", "initial subjective aim" and "final subjective aim", to take the fundamental components of the actual occasion into consideration. Here, the actual occasion stands out as a thing-in-itself, dissected away from the secondary considerations and refinements of Whitehead's concepts.

The primary components of the actual occasion are (i) prehension, (ii) integration, (iii) subjective aim, and (iv) objectification. Important ancillary notions are sense data and satisfaction. These components are closely related to the component processes of a sensorimotor control system: (i) sense, (ii) comparison, (iii) purpose, and (iv) action, respectively (Figure 2).

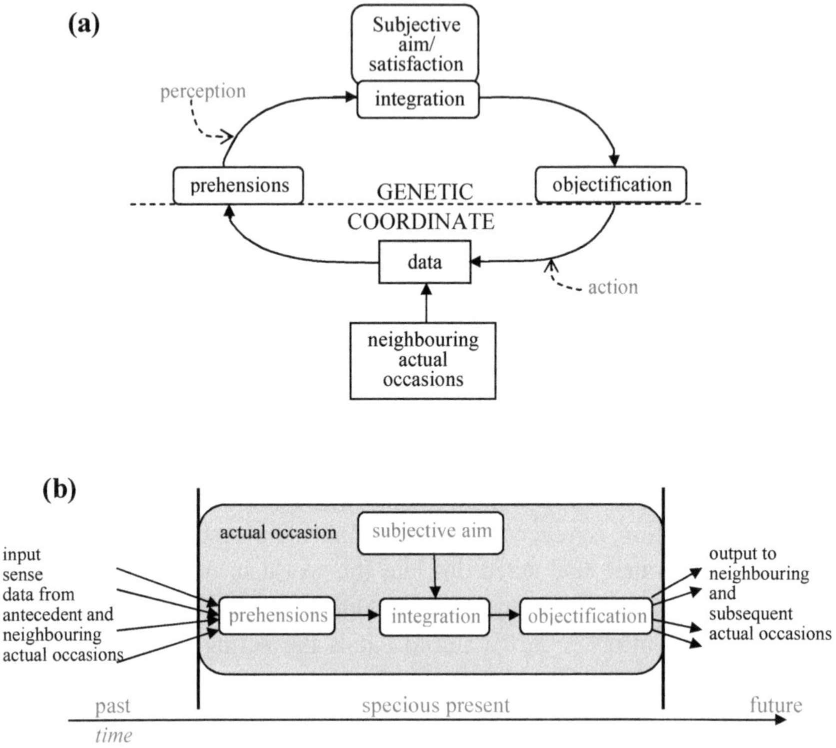

Figure 1: (a) Schematic of an ontological unit in action. The process is quasi-simultaneous but not strictly cyclical. Each component in the ontological unit affects the other, downstream components to yield a single unit driving toward its subjective purpose. Prehensions feed the integration and the subjective aim feeds both the prehensions and the integration. The difference between the control system (Figure 2) and the ontological unit, (a), is the sense of circularity. The ontological unit is strictly a linear process that reaches satisfaction through objectification, though it does maintain a cyclical nature though the acts of repeated integrations and re-integrations of prehensions.
(b) Schematic of the ontological unit in (a) linearized to time. Sense data is prehended and integrated until satisfaction is reached, guided by the subjective aim. The final act of the unit is objectification of itself into neighbouring and subsequent actual occasions, which yields both satisfaction in the unit itself and objectification with neighbouring and subsequent units.

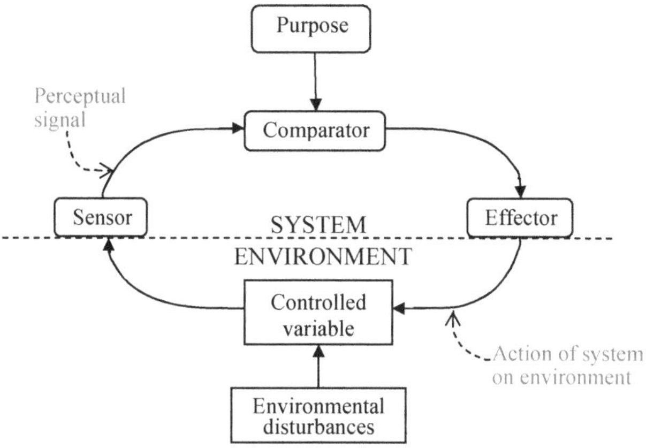

Figure 2: Schematic of an elementary control system in action. The process is quasi-simultaneous and cyclical, each component in the sensorimotor circuit affects the other, downstream components to yield a single unit driving toward its dictated purpose. Mechanical systems will continue indefinitely until the purpose changes or the system breaks down (adapted from Cziko 2000).

Whitehead's philosophy is based on process, and process is expressed through the movement of materials. It makes sense to map the component parts of the actual occasion onto the component parts of a sensorimotor control system; the component parts of the actual occasion and their arrangement fit the component part of the control system and their arrangement almost perfectly. I call this arrangement the "ontological unit" to differentiate it from Whitehead's more complex thinking about the actual occasion and to help us think along new lines of thought.

This unit forms the atomic unit of process in the philosophy of organism and it is this unit that lays at heart of Whitehead's ontology. By understanding this unit as an "atom" of existence, we place ourselves in a better position to understand how this part contributes to the greater experience of a whole in the living organism. How these atomic pieces fit together to explain the notion that "all things flow" (Whitehead 1979 [1929], 208 [317]) is an important question that can be addressed only once the atomic unit is known and understood. First, I would like to describe the components of the ontological unit.

ontological unit	sensorimotor unit
prehension	sense
integration	comparison
subjective aim	purpose
objectification	action

Table 1: Common components of Whitehead's ontological unit and Weiner's sensorimotor system.

Prehensions

A prehension is the first phase in the process of one unit relating to its antecedent occasion or to another unit. It forms the basis of what the unit knows to exist "outside" of itself, by way of the effect that the outside has on itself. The outside is replicated in a sense, through prehending. It is "sensing" and it is "perceiving" both. "A prehension reproduces in itself the general characteristics of an actual entity: it is referent to an external world, and in this sense will be said to have a 'vector character'; it involves emotion, and purpose, and valuation, and causation" (Whitehead 1979 [1929], 19 [29]). Thus, prehending has a mental pole as well as a physical pole, it is almost an entire actual occasion, but for the fact that it never reaches completion.

A prehension senses the objectified datum "whose relevance provokes the origination of this prehension; this datum is the prehended object" (Whitehead 1967, 176). Importantly, prehensions arise from the objectifications of other antecedent and neighbouring actual occasions.

Prehensions are a part of the complex ontological unit, which is itself an act of awareness. A prehension is a quasi-, or incomplete actual occasion. An actual occasion is itself composed of a multitude of prehensions. The notion of prehensions is "founded upon the doctrine that there are no concrete facts which are merely public, or merely private (…) The sole concrete facts, in terms of which actualities [units] can be analysed, are prehensions; and every prehension has its public side and its private side" (Whitehead 1979 [1929], 290 [444]).

The notion of prehending lies at the heart of the actual occasion and Whitehead's philosophy of organism. It embodies the dipolar nature of reality: both public and private, physical and mental.

"The facts of nature are the actualities; and the facts into which the actualities are divisible are their prehensions, with their public origins, their private forms, and their private aims. But the actualities are moments of passage into a novel stage of publicity; and the coordination of prehensions expresses the publicity of the world, so far as it can be considered in abstraction from private genesis. Prehensions have public careers, but they are born privately" (*ibid.*).

Prehensions are channelled through the actual occasion, embodied in it, and expressed outward in objectification. "Every prehension consists of three factors: (a) the 'subject' which is prehending, namely the actual entity in which that prehension is a concrete element; (b) the 'datum' which is prehended; (c) the 'subjective form' which is *how* that subject prehends the datum" (*ibid.* 23 [35]).

Integration

The integration of prehended sense data comes about by the interactions of the related parts of the unit. The totality of the integration is the totality of the interactions of that unit. Importantly, integration is guided by the subjective aim of the unit, which works to guide the form of the integration to yield the acquisition of its aim, its goal. The full scope of integration in an actual occasion is complex and involves notions within the philosophy of organism that I will not touch on here, namely the notions of eternal objects and the differences between physical and conceptual prehensions. For our purposes, integration can be sufficiently understood very much by what the term implies, the coming together of many aspects into one. In Whitehead's terminology, this is the "concrescence". Thus, prehensions are "involved in concrescent integrations, and terminate in a definite, complex unity of feeling" (*ibid.* 56 [89]).

Subjective Aim

Guiding the prehensive integrations to a terminal unity is the "subjective aim" of the occasion. The subjective aim is dependent on the subjective form of the occasion acting together with the integration of prehending sense data. The subjective aim guides the process of integration to "satisfaction", or completion, and is tied to the notion of eternal objects. The subjective aim acts through the integration by informing it and guiding the integrating process.

Whitehead further describes it as "the lure for feeling. This lure for feeling is the germ of mind" (*ibid.* 85 [131]). Prehensions "are bound together within each actuality by the subjective unity of aim which governs their allied genesis and their final concrescence" (*ibid.* 308 [470]). Subjective form can be taken to refer to the material stuff and the subjective aim guides the vector quality or directionality of the actual occasion.

Objectification

"The term 'objectification' refers to the particular mode in which the potentiality of one actual entity is realized in another actual entity" (*ibid.* 23 [34]). It is the acting out of the actual occasion into the world and in doing so satisfaction is obtained in making physically real the experience of the subject. Objectification is the public expression of the private experience made real and shared with others. It is, for all intents and purposes, expressed in movement.

The Ontological Unit

Sense data from other objectifying actual occasions are prehended, integrated according to the subjective aim, and acted out into objectification with other actual occasions. This is Whitehead's ontological unit. His thesis proposes the world is made of many interlinking and succeeding processual units. Both matter and mind are included: the material as objectification of process and the mind as prehension and integrations according to the subjective aim of the occasion. Thus, in the human form we prehend the world as subjects, integrate these sense data and act out into the world in objectification. This is who we are. It is clear that who we are is not a static thing, we are not simply our form, our bodily material, but we are the experience of that material living, and living is action. We are constantly perceiving, integrating, and acting out. Our actions are who we are.

2. Units of action: *Tau* in perception-action

2.1. Perception-action

Perceptual phenomena in biological movement is necessary for action. Actions are characterised by their goal-orientation, a feature that would not be possible without perceptual foresight (von Hofsten 2004). Perception is tied up in the very act of moving with purpose. Noë (2003) sums up the unity of perception-action, "Perception is not something that happens to us, it is something we do." Perception is active and bound up in living and is, to an evolutionary degree, dependent on one's capacities for action, whether or not these capacities – or *affordances* – are actually enacted by the physical stuff of the organism or not. Thus, to perceive is not merely to have sensations, it is to have sensations that one can understand and make use of in the context of one's organismic possibilities for action.

Perception is also largely amodal. The act of touching better illustrates perception than vision, because touch more obviously requires an active engagement with the world. Touch is tied up in movement, touch requires movement generated from within the organism, as does vision or any other modality of perception. Even when one is absolutely still, the visual field is "kept alive" through the saccades and micro-saccades of the eye. These small, rapid movements allow the animal to *feel* its ambient, much as a Braille reader *feels* a text. Thus, to perceive the world, one is intimately dependent on one's perceptual organs, which are active organs, be they fingers or toes or antennae. Furthermore, perception is not an exclusive process of the brain, it is a directed activity of the body as a whole and the nervous system is merely a convenient evolutionary solution to binding disparate experiences together (Sherrington, 1947). A fundamental criterion for perception is action.

2.2. Principles of animal movement

The living organism is a perceiving, moving entity that navigates its environment with purpose. How does it do this? How does an animal act in the world with purpose? This is the principal question that drives the multidisciplinary perception-action sciences. For a satisfactory answer, one must take into account an animal's ecological possibilities, its perceptual extension, its biomechanical abilities, its integrative nervous physiology, and its neuromuscular biology, all in conjunction with its spatiotemporal trajectory from the animal's past into its self-made future. The perception-action approach has roots in the early work of the perception psychologist, J. J. Gibson (1966, 1979), and his Soviet motor control contemporary, N. Bernstein (1967). A student of Gibson, D. N. Lee, summarises the principles of animal movements distilled from the earlier works. It is worth quoting these principles in full:

> "*Movements are always guided perceptually.* Animal movements are brought about by the interactive dynamics of muscular forces with external forces such as gravity and friction. External forces are not always predictable and so any movement may deviate from its intended course. Therefore any movement must be monitored perceptually to allow appropriate muscular adjustments to be made. For example, the articular proprioceptive systems, comprising sensors in the joints, muscle, and skin, are constantly active during all movements, and the vestibular systems are always active during movements of the head. These systems provide proprioceptive feedback to allow a movement to be adjusted over its course so that it maintains its intended course.
>
> *Movements are always guided intrinsically in relation with the environment.* An animal fashions movements to its purpose and so certain aspects of the movement must be guided intrinsically. The activity of the movement itself comes from within the animal, and thus the movement comes from within the animal to work with its biomechanical characteristics in conjunction with its external environmental characteristics.
>
> *Movements are always guided prospectively.* The animal has limited power available for making a movement. Therefore if it does not manage its power resources prospectively, it could end up not having sufficient power to complete a movement properly. This could have dire consequences if an animal runs out of braking power when trying to stop at a cliff edge, for example" (Lee 2005).

2.3. Units of goal-directed action

Many actions performed by living organisms are clear, goal-oriented ones. For example, a bird makes its final approach to land at its nest, a football player kicks the football, and a frog extends its tongue to catch a fly. These *single* goal-directed actions are composed of a set of smaller actions that contribute to and produce the larger single one. For example the bird requires wing beats to reach its nest, each wing beat is composed of an upward movement of the wing and a downward thrust, and each wing beat is adjusted over the course of the landing to produce an overall smooth approach to land. The football player requires many adjustments and re-adjustments of his muscles to produce the overall trajectory of the foot to the ball. The frog requires extending parts of the tongue together in a smooth manner for full extension. These actions are each composed of a set of nested smaller actions of their contributing muscle fibres and collections of muscle fibres that altogether produce one coherent, whole, and purposive action. They are thus complete action units enacted by the organism characterised by their goal-directedness and composed of nested smaller units of the subsystems of the organism.

Nestings of smaller actions within larger actions, and therefore collectives of smaller actions "giving rise to" larger actions, forms an hierarchical structure reminiscent of Laszlo's "natural hierarchy" (Laszlo, 1972). The suggestion is that each small component action of the larger, purposive action units made by the animal is itself an expression of a perception-action process made at a lower level, but following the same "rules" of perception-action cycles that the larger ones follow. In other words, they operate according to the same principles and can be considered to be individual units comparable to Whitehead's ontological units, *i.e.* actual occasions, with all the sense of autonomy, perceptual awareness, and acting out toward satisfaction that its constitution and environmental constraints permit.

In research on babies, von Hofsten (1979, 1991) found that goal-directed reaching movements performed by infants were composed of several phases of acceleration-deceleration. He termed these single acceleration-deceleration events "movement units". At birth, the infants' reach to grasp an object is composed of several movement units, several cycles of acceleration and deceleration. However, in adults the reaching to grasp is performed in one smooth movement unit, suggesting that a feature of developmental advance is the integration of multiple movement units into one coherent whole. Thus, in development many disparate units are brought together in one to form contributory, fluid wholes.

Goal-oriented displacements of parts of the body or of the whole body to obtain goals are our perfect, exemplar units of actions. These actions are clear and well defined by high spatiotemporal resolution motion capture technologies. Additionally, they are as true to our experience of life as they are clear to our observation and measurement. We enact them in experimental settings, driven by our volitional experience, and monitored closely by motion capture. The fact that they are goal-directed cannot be refuted. They are the units of motoric displacement toward unequivocal targets. There is no argument against teleology here: these are small, immanent teleologies enacted by the organism.

Using this basic framework for an understanding of perceptuomotor control in general, we can step beyond obvious displacement activity to examine other forms of animate movement and so generalise our understanding of goal-acquisition to include other modalities. One such example is mammalian sucking, which involves complex, neural control of the mouth, tongue, and pharynx musculatures. The result of these coordinated neuromuscular activities is to produce negative pressure in the oral cavity. In sucking, intra-oral suction pressure is "thing" effectively sensed and controlled. The regulation of the intra-oral pressure is directly coupled to the rate of the flow of the liquid being sucked, such as milk during a breast feed. The displacement in this case is not so obvious, it is the combined tensions and displacements of the oro-motor structures, but the goal is clear: to draw in liquid in a controlled manner. Studies on sucking have shown that they follow the same principle of perceptuomotor control that the limb displacements follow, namely sensing the rate of gap closure and controlling that change (Craig and Lee 1999).

2.4. General tau theory

How do animals perceive and move with purpose in their world? If we consider the principles of animal movement, that movements are always guided perceptually, intrinsically, and prospectively, then it follows that

(1) the guiding information must embrace the future, and that
(2) movement guidance must be simple, rapid, and reliable – and probably follows universal principles (Lee 2005).

The "information" that is being sensed and controlled in any organism must work prospectively to allow adequate extrapolation of the movement into the future. It must therefore have a temporal structure that extends it forward beyond the immediate present. For a natural system to do this, it must do so with simplicity and efficiency to work within the laws of parsimony.

Furthermore, the fact that animals with very small or with no nervous systems, such as insects or protozoa, move with a precision comparable to our own, suggests common underlying principles of movement guidance are prospective (Delafield-Butt, Pepping, McCaig, & Lee 2012).

General *tau* theory has been built on these principles of animal movement to account for the simple, rapid, and reliable perceptuomotor control of movements to goals observed in nature. The theory hinges on the notion that for any organism, it is *tau* information that is being sensed and controlled. *Tau* (from the Erasmian pronunciation of the Greek letter, τ, used to denote the variable mathematically) is an experienced variable of the "time to gap-closure" of any possible, prospective goal (*cf.* any affordance). It is an extraordinary informational variable, yet it may be one of the commonest features of animal perception (Delafield-Butt and Schögler 2007; Lee 2009). Typically in experimental situations, one measures the world with public units such as metres, grams, seconds, and so on. *Tau*, however, is a "measure" experienced by the organism, which we can infer through the mathematical analysis of public measures of its changing relations with the world of objects that it is interacting purposefully with.

Any one *tau* is therefore a perceptual measure of an affordance available to the animal. It is calculated as the "time to goal-acquisition at its current rate of closure", but is specifically the experience of one's path into future relations, a perceptual awareness made useful by directly enabling its content to be available for controlling one's changing relations (by coupling one *tau* with another, see below). It is biological perceptuomotor pragmatism, embodying all the necessary information in one effectual unit that cuts across all modalities of perceiving and acting.

A *tau* can be expressed mathematically as

$$\tau = x / \dot{x}, \qquad \text{(Eqn. 1)}$$

where x is the Cartesian coordinate distance to the goal and $\dot{x}$ is its first-order time derivative. Thus, in an experiment we can measure the distance-to-goal in metric units (*e.g.* metres) and the first-order time derivative in metric units per unit of time (*e.g.* metres per second). However, the unit of measure of the *tau* simply becomes a measure of time, because the Cartesian distances cancel each other out. This raises a very interesting issue, because while we can scientifically measure the phenomenon of time in units of seconds, the experiencing organism is not likely to be doing this. Rather, it is *experiencing its "time"* to closure, which is not the same as the public time that we measure with clocks, but is a unique perceptual measure generated within the organism and with its changing environment.

Tau perceptual information can be used for any modality of perception; it is theoretically a perceptual universal. We can think of it in terms of *changing intensities framed by goal-orientations*, where the intensity refers to any physical modality. Experimental evidence indicates *tau* information is perceived visually, acoustically through echolocation, through proprioceptive feedback, and through pressure in a number of different species tested (see Lee 2005, 2009 for a review).

Perceiving a *tau* is the first part of a perceptuomotor process. The second involves the useful manipulation of the *tau* to act prospectively in the world. The evidence indicates this is achieved by coupling one *tau* (*e.g.* the *tau* of gap '*p*') onto another *tau* (*e.g.* the *tau* of gap '*r*'), and keeping these *taus* in constant proportion so that

$$\tau_{(p)} = k\tau_{(r)} \qquad \text{(Eqn. 2).}$$

Thus, if gap p is the perceived gap between the interception of a hand to a ball and gap r is the perceived gap between the ball and the its interception with the hand, then by coupling gap p to gap r results in the hand reaching the ball at the interception point at the right time (*e.g.* Lee *et al.* 2001). Coupling *taus* ensures a stable relationship between the two, and ensures that they will physically meet. The coupling constant, k, determines the form of the changing relationship between the two *taus*, and so also the changing relationship between the objects that are creating the *taus* in the organism. By keeping k constant below a value of 1, the two objects will meet in a form with precise spatiotemporal dynamics. The value of k determines those dynamics (Lee 1998, Lee *et al.* 1999, 2001). The manner or form of the movement lies along a spectrum from "hard" to "soft", where a value of k between 0.5 and 1 produces a hard impact, a controlled crash as it were, and a value less than 0.5 produces a soft, gentle touch on goal-acquisition.

Interestingly, the k value is established at the beginning of the action. Over the course of the movement, the animal attempts to keep the k value constant. There is oscillation above and below this optimal value, akin to a tight-rope walker swaying left and right during his glide over the rope.

Over the course of each *tau* perceptuomotor action, there is feedback and integration, and correction and control, to maintain the main aim of the movement with its course through spacetime. The completion of the action is achieved by trying to keep its coupling constant constant.

In general *tau* theory there are two forms of coupling that are important. In the first case external objects may be coupled onto the movements of the animal or parts of the animal. While these are interesting from a motor control point-of-view, the more interesting mode of control for our purposes are the *internally-guided* controls. These movements originate from within the organism to act out into the world, not necessarily to intersect with an external object, but simply to act out. These movements are intrinsically guided.

The postulate put forward (Lee, Craig and Grealy 1999; Craig, Delay, Grealy and Lee 2000) suggests that these movements are coupled onto an "internal guide" that takes the form of either (1) the movement of a body under constant acceleration, or (2) the movement of a body under constant deceleration. These guide dynamics are equivalent to a body free-falling downward under gravity and a body losing momentum as it moves upward under gravity, respectively. I will focus on the former, mathematically expressed as

$$\tau_G = 0.5(t - T^2 / t) \qquad \text{(Eqn. 3)}$$

where t is the instantaneous present time and T is the total duration of the action. Internally-guided movements give an external spatiotemporal "contour" to the action, specifying the changing kinematic process required to produce a reliable pattern of change that enable goal-acquisition. In short, the internal guide provides an idealised kinematic distribution of accelerative then decelerative forces over its time course (T) to arrive efficiently and with controlled contact at its goal. There is a beginning, an integrative narrative, and an end. The work of the system acts out into the world to acquire a goal.

2.5. The process of an action unit

The perceptuomotor process of coupling the external expression of an action with its internal guide can be summed up diagrammatically (Figure 3), not only to conveniently describe its process, but also to draw on immediate parallels to the ontological unit. In any *tau*-guided perceptuomotor process there is an initiation toward a goal based on an initial subjective aim, there is the process of integrating and re-integrating sense data to achieve this aim, and there is the final completion of the act.

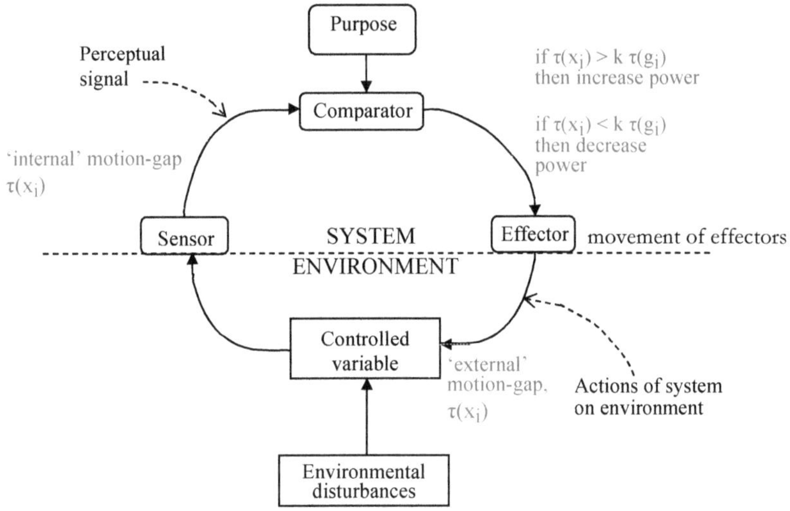

Figure 3: Schematic of a *tau* guide control system. The animal's purpose is to obtain x, a goal state in any modality. Acquisition of the goal is made efficiently by coupling the motion-gap, $\tau(x)$, onto an internal *tau* guide, $\tau(g)$. These taus are kept in constant ratio by adjusting the power delivered to the effectors. Physics ensure the internal and external variables, $\tau(x)$, are directly proportional, thus closing the internal gap also closes the external one. The quasi-circular process continues until the values of τ reach 0, at which point the goal is achieved and the process ceases.

These phases are as follows. The initiation is the establishment of a goal-orientation and its first impulse toward it. This is stated mathematically as the establishment of an initial coupling between two taus with a constant, *k*. As the process develops toward its goal with reiterative acts of integrating new sense data, the *k* value may shift from its initial ideal due to unforeseen forces, a buffet of wind or a slip of the foot, for example, veering the process off course slightly. The coupling constant is adjusted over to compensate and adapt to these changes, always in the present moment to achieve the subjective aim. The nature of the subjective aim of coupling taus together in constant ratio ensures the goal is acquired. The action then perishes, by which point new actions may have already arisen, taking the organism off into other futures.

2.6. Expressive Movements

If we extrapolate the experimental evidence to expand the scope of *tau* perceptuomotor control processes, we can consider that all human movements are performed in this same way. Expressive movements made in communication with others, or with objects in our environment, such as a piano during music-making, have a more immediate salience to our individual, felt experience, because we generate them to clearly express a felt quality within. They are all we can do to express ourselves as living, feeling entities. We do not have any other way of communicating our experiences with others except to move, to act. And though expressive movements of all kinds, such as through music, gesture, or talk, we live and share our lives with each other. In fact all movements, including the less affective and more directed ones described above, or indeed the movements of state changes of our autonomic neurophysiologies, are "expressive" insofar as they are an outward, physical expressions of an internal state. Movement generated or guided by an internal force is always an outward, physical expression of the internal system.

Analysis of emotionally salient "expressive" movements in dance and in music-making have been performed. They demonstrate the use of tau perceptuomotor control in the coordination of the complex movements of the diaphragm, chest, larynx, and facial muscles to arrange the oral and nasal cavities for acoustic reverberation during singing (Schögler, Pepping, & Lee 2008). Each of these components enact single actions with a clear beginning and a clear end – they aim for their target and acquire it in one smooth movement – that works in concert together with the other component actions to construct a larger, expressive action. Thus, they are a set of coordinated smaller actions made to obtain the overall composition of a smooth movement. An enormous multitude of smaller movements must always be coordinated and nested within the overarching single one.

Communication is expressed through changes in our musculature. Our whole bodies are acting in communication with other bodies to produce changes in posture, poise, intention, inflexion, and so on. Sound produced by the larynx is created in concert with body posture and modulated by changing dynamic acoustics of the oral cavities. The body is moving with the chest, larynx, throat, and the complex oral structures. Music and dance are refined, coordinated narratives of movements that produce articulations of thoughts and feelings. Similarly, consider movement in communication with intimate others. We move our hands, arms, chest, head, and facial muscles together with the many muscles of the larynx, chest and oral muscles to produce verbal language. Consider telling a story to family and friends. Movements say everything we are able to say about ourselves, about our inner experience of the world. Movements are connected intimately to perception, to mind.

One way to grasp *felt* experience in movement-making is to study spontaneously produced improvisations in paradigms such as music making. Here the felt experience of the musical narrative with all its emotional and perceptual depth is spontaneously created into the future by the musical participants. Perception and action are tied equally to the overt movements of communicative expression as they are to the more mundane actions such as kicking, reaching, or sucking, but the feeling accompanying them is richer in affective quality. Both are rich, mindful experiences framed by their immanent goals.

3. Discussion

3.1. Weaving taus to create complex movements / weaving units to create enduring form

A tau-guided perceptuomotor control action is a single action, it is an atomic unit of action with a beginning, a narrative of development and integration, and an end. However, most movements are continuous and fluid, as in the case of running, singing, and dancing. In these cases, the totality of the movement is made up of a collection of goal-directed movements arranged in an overlapping weave, much like music is an interwoven collection of changing notes.

The case of the sprinter demonstrates the principle of weaving *taus* to create complex movements (Lee 2005). The various musculatures of the sprinter's legs must work to perform a series of "contrapuntal", goal-directed movements to give circularity to each foot's movement. Each "voice" in the foot's movement is a single goal-directed movement: moving from its highest point to the ground, moving from its most extended point forward to its most extended point behind, moving from the ground to its highest point, and moving from behind the sprinter to in front of the sprinter. Analysis shows that each of these parts are separate goal-directed movements, yet they overlap and in doing so they form one continuous, fluid motion in the same way that a piece of contrapuntal music does.

The similarity with interwoven actual occasions is apparent. Each actual occasion is a goal-directed moment itself. Like each tau-guided perceptuomotor action, it is a discreet unit. And like complexes of tau-guided perceptuomotor actions, complexes of ontological units woven together form a unified "thing" that endures as a particular character, not unlike the sprinter being a particular kind of "thing" as she is sprinting. Ontological units can feed one to the next in a string of actual occasions, or they can feed into neighbouring actual occasions. As one perishes, it feeds into a growing other. The arrangement of these perishings and feedings is important for an understanding of how actual occasions are interwoven, and indeed if the parallel holds with actions, how individual action-units feed one to the other to form a coherent thing. Agar (1936, 1943) discussed this with some excellent foresight.

3.2. Atomicity and the Epochal Theory of Time

I have set out the parallels between Whitehead's ontological unit and perceptuomotor units of action. These two units, arising from different disciplines to explain different, yet comparable phenomena are remarkably similar in form and construction (see Figures 1b and 3). Yet, there is a major distinction between them that may be quite fundamentally revealing. Whitehead states that "the creature is extensive, but its act of becoming is not extensive" (Whitehead 1979 (1929), 69 [107]). If we take the "act of becoming" to be the process of achieving satisfaction in an ontological unit, then temporal extension is, according to Whitehead, not a part of that act. This is a point of contention, because to exist one must have duration (extension) through spacetime, a spatiotemporal presence. Whereas Whitehead seems to be saying that the process is not spatiotemporal.

However, this problem recalls a curious feature of *tau*: *tau* "reverses" time so that temporal extension is framed in goal-directed, "future" possibilities, and the process of the act of achieving an aim is to keep the relationship of two *taus*, two future-oriented subjective frames, in proportion. Thus, the theory suggests that we frame our perceptual existence on *taus* afforded us through our self-environment coupling, and as we approach a chosen *tau* percept we do so by keeping that *tau* in constant proportion to another *tau*, the internal "guide", say. For any *tau* perception-action process, time (the shared, public clock time) is "reversed" and structured by anticipation of possible goals. Many collections of these *tau*-times build themselves up within a human organism to be roughly comparable collectively to a clock time, though our actual experienced time is not like clock time at all (Pöppel 1994, Schögler 1999). Thus, in general tau theory "time" is a perceptual quality based on anticipated futures, some of which will be appropriated through prospective control by coupling anticipated futures together, by coupling *taus* together.

Like the actual occasions, discreet goal-oriented sensorimotor control processes are atomic action units that altogether compose the overall behaviour of an animal. Like actual occasions, they are made up of smaller contributing units of sensorimotor control processes. They are extensive, but they are not extensive because they are made of discreet parts themselves. The model is one of a "weaving" of units, where overlapping and inter-related sensorimotor processes work in harmony to produce the one overarching coordinated action that we measure, similar to how Whitehead describes the contribution of actual occasions to the enduring society of actual occasions,

"... in every act of becoming there is the becoming of something with temporal extension; but (...) the act itself is not extensive, in the sense that it is divisible into earlier and later acts of becoming which correspond to the extensive divisibility of what has become" (Whitehead 1979 (1929), 69 [107]).

4. Conclusion

Change is unceasing. Life is the control of that change into enduring patterns. Movement is change, all changes are movements. Life is the control of movements at atomic, molecular, cellular, cellular systems and whole organism levels. Living organisms appropriate masses and energies from their ambient into the constitution of their being, appropriated and directed there for *their* purposes. Living requires the continuous appropriation of these masses and energies into the constitution of living bodies and movement is both the displacement of these particles themselves, the displacement of the animal's body, its skeletomusculature, and the physiology and biochemistry occurring within the animal to make these movements possible. The process of appropriation into the living thing takes movement and involves movement. It takes movement of one kind and appropriates into the movement of another kind, its kind. All of nature in motion. The disparate movements of energies of the inanimate world are absorbed and channelled into the living organism where they become patterned and contained in precise spatiotemporal morphologies that we recognise as the living thing.

Material is always being ingested by an animal. Ingested complex molecules contribute directly to the physical substance of the living thing, making up the structure of its body as in the case of ingested proteins, or they give the energy required for the movement of the living thing. Ingested appropriation of physical substances forms the meat, the body of the animal; they are the solid, heavy substance of its existence.

Yet, another kind of material is also ingested, continuously. This material is the ambient stuff that make up our sensory world. It is the photons, diffuse molecules, molecular vibrations, and pressures of the world around us: the ambient energies of our environment. These too are appropriated in such a way that they become constant in their "appropriation" by the organism (Powers 1973). They become constant to the organism so that the organism's experience of them remains constant, relative to the chaos of those energies in the world around it. It is this latter form of appropriation that I have talked about here, it is the control of the movement of the animal to keep its world within its sphere of possible influence, within its particular vital realm of experience, which is its constant ambient, its constant habitat, its safe place in which to live.

REFERENCES

Agar, W. (1943). *A Contribution to the Theory of the Living Organism*. Melbourne and London: Melbourne and Oxford University Press.

—— (1936). "Whitehead's Philosophy of Organism: an Introduction for Biologists". In: *The Quarterly Review of Biology*, 11, pp. 16-34.

Bernstein, N. (1967). *The Co-Ordination and Regulation of Movement*. Oxford: Pergamon.

Craig, C.; Delay, D.; Grealy, M.; Lee, D. (2000). "Guiding the Swing in Golf Putting". In: *Nature*, 405, pp. 295-296.

Craig, C.; Lee, D. (1999). "Neonatal Control of Nutritive Sucking Pressure: Evidence for an Intrinsic Tau-Guide". In: *Experimental Brain Research*, 124, pp.371-382.

Cziko, G. (2000). *The Things We Do: Using the Lessons of Bernard and Darwin to Understand the What, How, and Why of our Behavior*. Boston: MIT Press.

Delafield-Butt, J. ; Pepping, G.; McCaig, C.; Lee, D. (2012). "Prospective Guidance in a Free-Swimming Cell". In: *Biological Cybernetics*, 106, pp. 283-293.

Delafield-Butt, J.; Schögler, B. (2007). "The Ubiquitous Nature of Tau". In: Pepping and Grealy (Eds.). *Closing the Gap: The Scientific Writings of David N. Lee*. New York: Erlbaum.

Gibson, J. (1979). *The Ecological Approach to Visual Perception*. Boston: Houghton Mifflin.

—— (1966). *The Senses Considered as Perceptual Systems*. Boston: Houghton Mifflin.

Hofsten, C. von (2004). "An Action Perspective on Motor Development". In: *Trends in Cognitive Sciences*, 8, pp. 266-272.

Laszlo, E. (1972). *Introduction to Systems Philosophy: Toward a New Paradigm of Contemporary Thought*. New York: Gordon and Breach.

Lee, D. (2009). "General Tau Theory: Evolution to Date". In: *Perception*, 38, pp. 837-858.

—— (2005). "Tau in Action and Development". In: Rieser, J.; Lockman, J. and Nelson, C (eds.). *Action as an Organizer of Learning and Development*. New Jersey: Erlbaum.

——(1998). "Guiding Movement by Coupling *Taus*". In: *Ecological Psychology*, 10, pp. 221-250.

Lee, D.; Georgopoulos, A.; Clark, M.; Craig, C.; Port, N. (2001). "Guiding contact by coupling the *taus* of gaps". In: *Experimental Brain Research*, 139, pp.151-159.

Lee, D.; Craig, C.; Grealy, M. (1999). "Sensory and intrinsic coordination of movement". In: *Proceedings of the Royal Society of London* B, 266, pp. 2029-2035.

Noë, A. (2005). *Action in Perception*. Cambridge, Mass.: MIT Press.

Pöppel, E. (1994). "Temporal mechanisms in perception". In: *International Review of Neurobiology*, 37, pp. 185-202.

Powers, W. (1973). "Feedback - Beyond Behaviorism". In: *Science*, 179, pp. 351-356.

Reed, E. (1996). *Encountering the World: Toward an Ecological Psychology*. Oxford: Oxford University Press.

Schögler, B. (1999). "Studying Temporal Coordination in Jazz Duets". In: *Musicae Scientiae, Special Issue, 1999–2000: Rhythms, Musical Narrative, and the Origins of Human Communication*, pp. 75-89.

Sherrington, C. (1947). *The Integrative Action of the Nervous System* (2nd ed.). Cambridge: Cambridge University Press.

——(1991). "Structuring of Early Reaching Movements: a Longitudinal Study". In: *Journal of Motor Behaviour*, 23, pp. 280-292.

——(1979). "Development of Visually Directed Reaching: the Approach Phase". In: *Journal of Human Movement Studies*, 5, pp. 160-178.

Whitehead, A. N. (1979). *Process and Reality.* New York: Free Press/[(1929). New York: Macmillan].

——(1967). *Adventures of Ideas*. New York: Free Press.

Life in the Interstices: Systems Biology and Process Thought

JOSEPH E. EARLEY, SR.

Insights gained by contemporary biology should interest philosophers, whether or not they agree with E. O. Wilson's statement:

> "[Biology] has become foremost [among the sciences] in relevance to the central questions of philosophy, aiming to explain the nature of mind and reality and the meaning of life" (2006, 106).

Clearly, philosophy of science finds significance in current biology. Vigorous discussions of levels of selection, mechanistic explanation, and other technical topics are now in progress. However, Wilson suggests (correctly, in my view) that philosophers other than philosophers of science (e.g., metaphysicians) also should take insights gained in recent biological research into account. In particular, investigation of how interactions between components of biological entities lead to the behaviors and properties of those entities – the currently active field of "systems biology" (Klipp 2005) – will repay philosophical attention. As eminent bioinorganic chemist R. J. P. Williams puts it:

> "Studies of biological sciences can be approached in two ways: reductively, as in molecular biology, or holistically, as in systems biology. [...] The future lies with the second [way] as the first is nearing completion" (2005).

The research program that grew into contemporary systems biology developed in the decades just before the middle of the 20th century. The main features of the quantum interpretation of microphysics and Whitehead's "philosophy of organism" also matured in the same period. C. H. Waddington (1959), a pioneer of systems biology, recognized that Whitehead's philosophy was directly relevant to what he called "theoretical biology". Many authors have called attention to connections between quantum-mechanics and Whitehead's philosophy. The possible relevance of both quantum considerations and process thought to the philosophy of mind has often been discussed (e.g., Eastman 2004; Gunter, this volume). However, as Atmanspacher remarked:

> "It turns out that the implementation of events in Whitehead's sense into quantum theory is everything else than straightforward. The even more difficult inclusion of mental time remains mostly unaddressed" (2006).

This paper reviews Whitehead's doctrine that indeterminacy is essential for both life and mind, and suggests that necessary indeterminacy is more likely to arise from networks of relationships (such as those considered by systems biology) rather than from quantum-mechanical features of microphysics. I also sketch a neo-Whiteheadian metaphysical approach (Process Structural Realism, PSR) that can incorporate both the findings and the spirit of systems biology.

I

Whitehead rejected dualistic metaphysical approaches that have long history, widespread influence, and vigorous present defenders (e.g., van Imwagen 2002). But Whitehead's own notions of life and consciousness seem to be subject to interpretations that tend to shade toward dualism. In this respect, an especially problematic feature of Whitehead's system is his doctrine that:

> "[…] life is a characteristic of 'empty space' and not of space 'occupied' by any corpuscular society. […] Life lurks in the interstices of each living cell, and in the interstices of the brain" ([1929] 1978, 105-106).

To the unwary, this might seem to suggest that some microscopic entity (perhaps a *res vivens*) jumps from cavity to cavity inside each biological cell, and some kind of homunculus (*res cogitans*) cavorts inside the skull of each conscious human individual. To avoid these notions, so foreign to the spirit of Whitehead's project, one must recall that Whitehead replaced the notion of 'absolute' space and time that Newton had laid out in his

Scholium (and which still provides an unexamined basis for much thinking, even among philosophers) with the concept of "the extensive continuum" – a plenum of possibilities. For each actual entity in the course of its concrescence, the infinite potentiality of the extensive continuum is reduced to a definite finite actuality by antecedent actual occasions that constrain the concrescent entity in its coming-to-be.

> "Continuity concerns what is potential; whereas actuality is incurably atomic ([1929] 1978, 61) [...]. An extensive continuum is a complex of entities united by the various allied relationships of whole to part, and of overlapping so as to possess common parts, and of contact, and of other relationships derived from these primary relationships. The notion of 'continuum' involves both the property of indefinite divisibility and the property of unbounded extension ([1929] 1978, 66). [...] Actual entities atomize the extensive continuum. This continuum is in itself merely the potential for division; an actual entity effects this division" ([1929] 1978, 67).

For living entities, however, constraints arising from antecedent actualities are not totally effective.

> "The emergence of life is [...] a bid for freedom on the part of organisms, a bid for a certain independence of individuality with self-interest and activities not to be construed purely in terms of environmental obligations" ([1927] 1985, 65).

It is essential to Whitehead's notion of life that some flexibility always remains to be resolved by the self-creative decision of each living organism. Isabelle Stengers (2008) emphasizes the importance of "non-conformal propositions" (possibilities not fully specified by past history and a particular environment) in achieving the novelty in concrescence that life requires. I now suggest that what Whitehead calls "'empty' space" should be considered to be a metaphorical space of indeterminacy, rather than some gap in extension in Newtonian absolute space. In this view, what Whitehead refers to as "interstices in the brain" would be real possibilities for mental coherence that are not yet realized – and therefore are available for prehension in the course of self-creation of a conscious superject. Those interstices should not be regarded as otherwise-unoccupied spatial volumes within a human skull but rather as unrealized possibilities for coherent neural activity.

> "[...] It seems as though the last delicacies of feeling require some element of novelty to relieve their massive inheritance from bygone system. Order is not sufficient. What is required, is something much more complex. It is order entering upon novelty; so that the massiveness of order does not degenerate into mere repetition; and so that the novelty is always reflected upon a background of system. [...] It is by reason of the body, with its miracle of order, that the treasures of the past environment are poured into the living occasion. The final percipient route of occasions is perhaps some thread of happenings wandering in 'empty' space amid the interstices of the brain" ([1929] 1978, 339).

Whitehead's doctrine of "order entering on novelty" may be regarded as a direct anticipation of an important concept of systems biology – that evolutionary systems generally tend towards "the edge of chaos" (Kauffman 1993).[1] It has been found that evolutionary systems generally tend towards an organized state that differs from disorderly regimes only in relatively small variations. Each life-form that is not stuck in an evolutionary back-

[1] It is usual to represent the state of any evolutionary system as a point on a multidimensional 'fitness landscape'. (This is based on a map that has one dimension for each of the many factors that influence the reproductive success – the "fitness" – of that system. The fitness of a system is represented by a height above the plane of such a map.) Such a landscape is analogous to a relief model of a mountain range. Usually, such fitness landscapes are 'alpine' – with sharp peaks, deep ravines, and few relatively even plateaus. More-fit systems necessarily out-reproduce less-fit ones (reproductive success under the conditions that prevail is identical with fitness). Therefore, evolutionary development by gradual incremental alteration always leads to higher fitness. That is, evolutionary systems necessarily climb to higher altitudes on a fitness landscape; they always move up-hill. Therefore, the eventual fitness of each system will be limited by the height of the 'local peak' on the side of which that system started, even if the same landscape contains peaks that correspond to much higher fitness. Those higher peaks will not be accessible to a system that has already moved to the top of the hill on which it started. Downhill migration on an unchanging fitness landscape is never possible for evolutionary systems, but there is no way to reach an alpine peak from a distant lowland by surface travel without sometimes going down a slope. In order to escape evolutionary dead-ends (becoming stranded on low fitness peaks), biological systems need mechanisms that lead to non-incremental (discontinuous) change. This sketch assumes that fitness landscapes are constant, i.e., not influenced by changes in the environment or variations in other species. Closely-related conclusions can be reached without this unrealistic simplifying assumption.

water[2] has two characteristics. Every successful species has properties that reproduce successfully from generation to generation. Each type of organism also has some features that result in occasional large changes. Most such major alterations do not actually lead to higher fitness – and therefore have no long-term consequences. However, from time to time, such a jump will lead to significantly higher fitness – an evolutionary advance will have occurred. This coupling of characteristics (general conservatism and occasional adventurousness) corresponds to a situation where order prevails, but which is similar to a disordered (chaotic) regime. A variety of evidence shows that evolutionary systems generally tend to such "edge-of chaos" states. Whitehead's notion of the essential connection between freedom and life ("order entering on novelty") anticipates this recently-developed conclusion of systems biology.

II

Since biological systems are made up of components that follow physical and chemical regularities, how could the indeterminacy that Whitehead's view requires arise? Many authors have suggested that such indeterminacy may arise from quantum-mechanical features of microphysics. Penrose (1999) and Hameroff (2002) have provided an unusually detailed model of one way that sub-microscopic quantum-mechanical phenomena might be relevant to human mental functioning. They propose that the network of micro-tubules that exist within each individual neuron may provide their contents sufficient isolation from their surroundings to permit "quantum superimposed mass movements which are well insulated from their environment. It may well be that within the tubes there is some kind of large scale quantum coherent activity, somewhat like a computer" (Penrose 1999, 131-132). A small-scale example of the type of coordinated mass motion that this model requires has been reported to account for rapid synchronization of remote active sites in a specific enzyme (Frank 2004). In this case, the surface of a catalytic protein molecule includes a region in

[2] The lamprey seems to have reached such an evolutionary dead end. Fossils indicate that ancient forms were indistinguishable from present lampreys.

which amino-acid residues are so organized that simultaneous shifts of protons occur between neighboring atomic centers along an extended line of such centers (a "proton wire"). This results in effective transfer of a hydrogen nucleus to a quite remote site in a remarkably short time. Identification of such coordinated movement of atomic nuclei indicates that the Penrose-Hameroff model is not totally impossible, but also shows the tremendous degree of self-organization that would be required within microscopic tubules for that mechanism for quantum neuroscience to be effective. The properties of non-locality, quantization, and computational power that are sought in quantum neuroscience are certainly available more readily by other means.

III

Generally, the state of any natural system depends on relevant boundary conditions. When we observe two of a given kind with different properties (say one object has a blue color, and another otherwise identical thing is red) we often suppose that some difference in conditions accounts for the difference in properties. Koutroufinis (this volume) points out that "bistable systems" are common in systems biology, and are philosophically interesting. In each such case, objects of a single type have quite different properties (such as being blue or being red) under conditions that are *identical* in all particulars. It turns out that what differs between the two states is the *history* of the systems. By *what means* identical conditions are arrived at determines the properties of each situation. Otherwise identical conditions may correspond to two quite different property-sets depending on past history. This remarkable situation is not some peculiarity of living organisms. Similar behavior (called 'hysteresis') is encountered in chemical and physical systems that are only moderately complicated – and certainly are much simpler than any biological system. Geissler (1981) described a rather simple chemical system that illustrates main features of bistability. Two solutions are pumped into the lower parts of a reaction chamber that contains a stirring device (a "continuously stirred tank reactor", CSTR). The solution thus produced makes its way to the top of the CSTR chamber and then exits through an overflow tube. Suppose the experiment starts with the pump operating slowly (say at a rate of 0.1, in appropriate units). Under those conditions, the solution in the chamber (and in the exit stream) is blue. If the pump is gradually speeded up, the solution in the cell remains blue until a rather high pump rate (say, of 9 units) is reached, then the solution rapidly becomes red. The transition pump-rate is quite precisely defined. In a typical experiment, the solution remains blue at a pump-rate of 8.999 units, but is definitely red at a pump-rate of 9.000 units. What is remarkable about this system is that when the experiment starts with a high pump-rate of 10 units (the solution is definitely red) and the pump-rate is gradually lowered, the solution in the reaction chamber *remains red* long after the pump-rate is lowered well below 9 units. The transit from red to blue occurs only at a quite low value of pump rate, say 1.0 units. (The solution will remain red indefinitely if the pump-rate is set at 1.001 units.) For the extensive region of pump-rate variation between 1.000 and 8.999 units the solution in the exit-stream and in the chamber may be *either blue or red*, depending on the past history of the system!

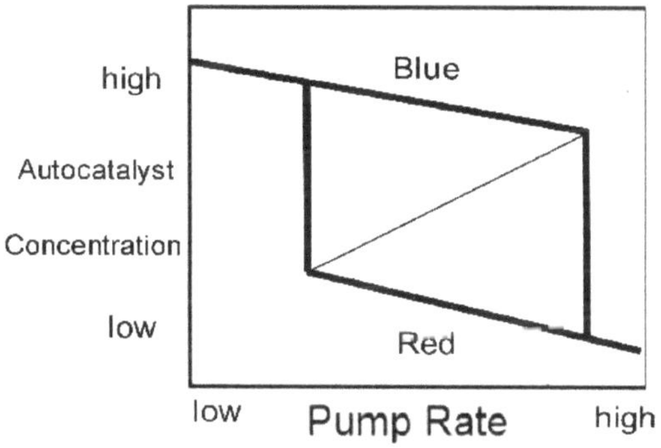

Fig. 1: Schematic diagram of the behavior of a bistable chemical system in a continuously-stirred tank reactor. At high pump-rates, the system has low concentration of autocatalyst and a red color. At low pump-rates, the system has high autocatalyst concentration and a blue color. At intermediate pump-rates, the system may be either red or blue, depending on whether the intermediate pump-rate was reached by increase or decrease.

The chemistry involved in this behavior is well understood (e.g., Scott 1994, 53-58). When chemical reactions involve a process that gets progressively faster as it proceeds (autocatalysis) and some second process slowly destroys some necessary component of the first reaction, then there may be *two* stable non-equilibrium steady states (labeled blue and red in the figure) as well as an unrealizable unstable steady state (thin diagonal line in the figure). At a constant pump rate, a system that exists in either stable steady state will stay in that state unless some external factor takes the system beyond the unstable state.[3] What accounts for this behavior is the concentration of one especially important chemical, called "the autocatalyst" (low in the red state, high in the blue). The rate of formation of this autocatalyst is slow, but once its concentration exceeds a critical value (the autocatalyst concentration characteristic of the unstable state) the color of the solution promptly changes. When the pump operates at a high rate, the residence time of the solution inside the reactor is not sufficient for enough autocatalyst to be generated to reach the critical concentration value. If the pump slows, reaction solutions remain in the reactor longer and more autocatalyst is produced. The transition of color occurs when autocatalyst concentration exceeds the critical value. When the pump is operating slowly, ample autocatalyst is generated and the system stays in the blue steady state. If the pump is speeded up from a low value, the second process may destroy more autocatalyst than is produced, and the autocatalyst concentration declines. Eventually, a pump-rate is reached at which autocatalyst concentration falls below the critical value and the blue to red transition occurs. The pump-rate at which this blue to red change occurs with pump-rate increasing is significantly higher than the pump-rate that corresponded to the red to blue change in the experiment with pump-rate decreasing.

A question that arises in all bistable situations is: How is one state or the other to be reached?[4] This question is related the medieval conundrum known as "Buridan's Ass". Does a hungry donkey starve when placed precisely midway between two identical bales of hay?[5] On what basis do biological systems, and individual biological organisms, resolve ambiguities?

[3] If a perturbation large enough to take the system beyond the unstable state were to occur, then the system would change to the alternate stable condition (from blue to red or the reverse).

[4] As the ancients said: "Which road leads to Delphi?"

[5] There is little comfort in the advice attributed to Yogi Berra: "When you come to a fork in the road, take it!"

By what means do natural systems settle on one of two alternative states (blue or red, say) in situations of bistability – when either of two states is compatible with the underlying conditions of the system? One approach to this question is to assume that if there is an "action," there must be an "actor" – every incident of apparent agency may be considered to imply the existence of an agent. On this view, if a selection is made, someone or something must have decided. This view is consistent with features of human mentality that have long evolutionary history. The harm an animal suffers in falsely identifying an agent that does not in fact exist is generally much less that the evil that would result if clues indicating a present predator were ignored. False positives are less risky than incorrect negatives.

An alternative approach to the question of how systems wind up on one of two possible results builds on Whitehead's observation:

> "However we fix a determinate entity, there is always a narrower determination of something which is presupposed in our first choice. Also there is always a wider determination into which our first choice fades by transition beyond itself" ([1925] 1967, 93).

Clearly, biological organisms are composed of subsystems (e.g., digestive systems, organs, tissues, cells, molecules…) that have greater or lesser degrees of integrity. Just as clearly, each biological organism is a component of larger units (breeding groups, broods, swarms, local ecologies, regional ecosystems…). Each of these persistent coherences, at both smaller and larger levels of size, is characterized by "homeostasis" – intricate balance of dynamic processes that maintains approximate constancy of overall properties under a range of environmental conditions. Further, each biological organism is the outcome of a long historical process of development; each species results from millennia of evolution. At each of the myriad stages of that evolutionary history, and at each of the many critical points of growth and development of each individual organism, units (of any and every level) that were not able to sustain homeostasis under whatever conditions actually prevailed failed to persist and/or to leave progeny.

Each biological network of processes involves features that are the functional equivalent of switches – variation in external conditions or of internal state gives rise to transitions of processes from one dynamic condition to another – from "off" to "on" or the reverse. Normally, such biological switching involves bistable systems. What may appear to be a basic ambiguity or indetermination of the state of an entity at one level may well be a feature that is intrinsic to a control network at another level. Bistable systems are relatively simple examples of coherent networks of dynamic processes. Those same bistable systems are also integral parts of more complex and larger dynamic networks. The characteristics and requirements of inclusive networks must be taken into account in understanding the factors that determine the state of a specific bistable system at a particular time. Chemists and other scientists can easily generate models of bistable systems that are largely independent of their environments. In natural systems such as those studied in systems biology, bistable systems (themselves networks of dynamic process of less-inclusive coherences) are quite generally functional parts of more-inclusive entities. Those larger networks must be taken into account in any attempt to describe why a particular bistable network functions as it does.

When the function of a network is unchanged over a range of values of the parameters that characterize that network, the network is said to be "robust" (Wagner 2005). In the CSTR experiment described above, quite similar results are obtained with solutions that are only approximately the same in concentration.[6] Another kind of robustness can be seen in "neural networks" – combinations of computing units have some similarities to how neurons may be connected in the human brain (Cattell 2006, 55-70). In such cases, the configuration in which units are connected is more significant than the properties of the individual units. Robustness is more the rule that the exception in the many classes of networks described in systems biology

Persistence of an inclusive ("upper-level") network requires that included ("lower-level") networks continue to function within certain tolerances. Lower level networks may be robust enough to have large degrees of flexibility – considerable lower-level variation may be undetectable in upper-level functioning. There may well be room for considerable variation among the details characteristic of the lower-level system. The robustness of networks can provide the 'empty' space and interstices (indeterminacy) that

[6] This experiment is sufficiently robust that "bartender's accuracy" suffices.

Whitehead's notions both of life and of consciousness require. Reasoning that involves network stability as a central concept is not as familiar as is discourse in terms of substantial things. Behavior of networks of processes displays strange and counterintuitive behavior, reminiscent of the problems and paradoxes that engaged early students of quantum mechanics. It seems much more likely that the properties of networks of dynamic relationship give rise to both mental and microphysical phenomena, rather than that mental phenomena result from some peculiarities of microphysics.

Understanding of complex networks has advanced rapidly in recent years, aided by progress in computer technology and also by advances in mathematics (e.g., Klipp 2005). This progress in nonlinear dynamics has contributed to major conceptual shifts in several scholarly fields. For instance, in economics, the myth of *homo economicus* – a fully rational agent who acts on the basis of complete information and pre-established values (utilities) – is being replaced by a view that envisions economic agents as *emerging* from dynamics of strategic interactions conducted under uncertainty (Bowles 2004, Beinhocker 2006). Philosophical understanding of progress in systems biology will also require attention to what may seem to be arcane details of nonlinear dynamics.

IV

The long-running discussion of scientific realism seems to have entered a structuralist phase that seems quite congenial to process philosophy. Elsewhere (Earley 2006a, 2008a, 2008b) I have developed aspects of an ontological approach that seems capable of dealing with concepts, methods, and results of systems biology and which also shares the basic intellectual thrust of research in that discipline. The main point of this approach is that when a group of processes achieves such closure that a set of states of affairs recurs continually, then the effect of that coherence on the world differs from what would occur in the absence of that closure. Such altered effectiveness is an attribute of the system as a whole, and would have consequences. This indicates that the network of processes, as a unit, has ontological significance. Whenever a network of processes generates continual return to a limited set of states of affairs, the system may function as a "whole" – with respect to appropriate interaction partners. The balance achieved by the processes provides the form of definiteness of a unified agent. The causal powers of such coherent aggregates are indeed just the powers of the "constituents acting in concert" (Merricks 2003). However, the components act in concert in the specific way they do *only because of* their inclusion in the closed set of interactions that defines the coherence. This renders the causal powers of the coherence defined by that closure non-redundant, and hence the coherence, as a unit, is ontologically significant. The form of definiteness that provides internal coherence also grounds external efficacy of the societal aggregation. The closure is a *structural* feature of the coherence – possibly, but not necessarily, apparent in spatial structuring. One can show (Earley 2006b)[7] that every such coherence is the representation of a mathematical "group" or "semi-group." What is fundamental is achievement of effective coherence – the level of size on which that achievement occurs is irrelevant. Combinations of processes produce effects that are not simply attributable to the constituents. Whenever that efficacy is relevant,[8] non-redundant causality warrants recognition of those coherences as ontologically significant. This ontology is a variety of structural realism – related to Ontological Structural Realism (OSR) (French 2003), but it is also a kind of process philosophy. The designation "Process Structural Realism" (PSR) seems appropriate. This approach can provide a unified account that includes quantum microphysics, systems biology, and the philosophy of organism – without reducing any of these to another.

References

Atmanspacher, H.; Primas, H. (2006). "Pauli's Ideas on Mind and Matter in the Context of Contemporary Science". In: *Journal of Consciousness Studies*, 13, pp. 5-50.

Beinhocker, E. (2006). *The Origin of Wealth: Evolution, Complexity and the Radical Remaking of Economics.* Boston: Harvard Business School Press.

---

[7] This involves Cayley's theorem.

[8] Whether or not coherence is ontologically significant depends on the detailed characteristics of entities with which that coherence interacts.

Bowles, S. (2004). *Microeconomics: Behavior, Institutions and Evolution*. Princeton: Princeton University Press.

Cattell, R. (2006). *An Introduction to Mind, Consciousness and Language*. London: Continuum.

Earley, J. (2008a). "Ontologically Significant Aggregation: Process Structural Realism (PSR)". In: Weber, M. (ed.). *The Handbook of Whiteheadian Process Thought*, Vol. 2. Frankfurt: Ontos Verlag, pp. 179-192.

—— (2008b). "Process Structural Realism, Instance Ontology, and Societal Order". In: Riffert, F. and Sander, H.-J. (eds.). *Researching with Whitehead: System and Adventure*. Freiburg, Munich: Alber, pp. 189-211.

—— (2006a). "Chemical 'Substances' That Are Not 'Chemical Substances'". In: *Philosophy of Science*, 73 (December), pp. 841-852.

—— (2006b). "Some Philosophical Implications of Chemical Symmetry". In: Baird, D. and Scerri, E. (eds.). *Philosophy of Chemistry: Synthesis of a New Discipline*. Dordrecht: Springer, pp. 207-220.

—— (2003a). "Constraints on the Origin of Coherence in Far-from-Equilibrium Chemical Systems". In: Eastman, T. E. and Keeton, H. (eds.). *Physics and Whitehead: Quantum, Process and Experience*. Albany: State University of New York Press, pp. 63-73.

—— (2003b). "How Dynamic Aggregates May Achieve Effective Integration". In: *Advances in Complex Systems*, 6, pp. 115-126.

Eastman, T.; Keeton, H. (eds.) (2003). *Physics and Whitehead: Quantum, Process and Experience*. Albany: State University of New York Press.

Frank, R.; Titman, C.; Pratap, J.; Luisi, Ben F.; Perham, R. (2004). "A Molecular Switch and Proton Wire Synchronize the Active Sites Thiamine Enzymes". In: *Science*, (29 October) 306, pp. 872-876.

French, S.; Ladyman, J. (2003). "Remodeling Structural Realism: Quantum Physics and the Metaphysics of Structure". In: *Synthese*, 136, pp 31-56.

Geissler, W.; Kedmar, B. (1981). "Bistability of the Oxidation of Cerous Ion by Bromate in a Stirred Flow Reactor". In: *J. Phys. Chem.*, 85, pp. 908-914.

Hameroff, S.; Nip, A.; Porter, M.; Tuszynski, J. (2002). "Conduction Pathways in Microtubules, Biological Quantum Computation and Consciousness". In: *BioSystems*, 64, pp. 149-168.

Gunter, P. (this volume).

Inwagen, P. van (2002). *Metaphysics*, Boulder, CO: Westview.

Kauffman, S. (1993). *The Origins of Order: Self-Organization and Selection in Evolution*. New York: Oxford University Press.

Klipp, E. et al. (2005). *Systems Biology in Practice: Concepts, Implementation and Application*. Weinheim: Wiley-VCH.

Koutroufinis, S. (this volume).

Merricks, T. (2003). “Précis of Objects and Persons”. In: *Philosophy and Phenomenological Research*, 67(3), pp. 700-703.
Penrose, R. (1999). *The Large, the Small and the Human Mind*. Cambridge: Cambridge University Press.
Riffert, F.; Sander, H. (eds.) (2008). *Researching with Whitehead: System and Adventure*. (Essays in Honor of John B. Cobb) Freiburg, Munich: Alber.
Scott, S. (1994). *Oscillations, Waves and Chaos in Chemical Kinetics*. Oxford: Oxford University Press.
Stengers, I. (2008). “Achieving Coherence. The Importance of Whiteheads’s 6th Category of Existence”. In: Riffert, F. and Sander, H.-J. (eds.). *Researching with Whitehead: System and Adventure*. Freiburg, Munich: Alber, pp. 59-79.
Waddington, C. (1959). *Biological Organization Cellular and Sub-Cellular: Proceedings of a Symposium*. London: Pergamon Press.
Wagner, A. (2005). “Circuit Topology and the Evolution of Robustness in Two-Gene Circadian Oscillators”. In: *Proceedings of the National Academy of Science (USA)*, 102(33), pp. 11775-11780.
Whitehead, A. ([1933] 1967). *Adventures of Ideas*. New York: The Free Press.
—— ([1929] 1978). *Process and Reality* (Corrected Edition; David Griffin and Donald Sherburne, eds). New York: The Free Press.
—— ([1927]1985). *Symbolism: Its Meaning and Effect*. New York, Fordham: University Press.
—— ([1925] 1967). *Science and the Modern World*. New York: The Free Press.
Williams, R. (2005). “Molecular and Thermodynamic Bioenergetics”. In: *Biochem. Soc. Trans.*, 33, pp. 825-828.
Wilson, E. (2006). *The Creation: An Appeal to Save Life on Earth*. New York: Norton.

Quantum Biology: A Live Option

PETE A.Y. GUNTER

Alfred North Whitehead once remarked that a society needs people for the locomotion of ideas as badly as it needs people for the locomotion of freight.[1] The present essay is written in precisely this spirit. The ideas to be presented here are not original to the author. But they are probably not familiar to the readers of this volume. They will be locomoted here not only because they are interesting, but because they are potentially of great importance. Should they come to dominate biology our notions of life, evolution, and of living things would be significantly, perhaps radically, transformed.

I would like, before starting, to draw attention to an article by this author in the first issue of *The Pluralist*: "Darwinism: Six Scientific Alternatives". This article forms the context of the present essay. It sketches a group of scientific ideas at odds with ultradarwin orthodoxy represented by Richard Dawkins, Francis Crick, and Daniel Dennett: nonlinear evolution, thermodynamic theories of evolution, neolamarckian evolution, Baldwinian ideas, the gene capture hypothesis (symbiogenesis), and quantum biology. The present essay, unfortunately, can deal only with the last of these.

There are good reasons why the possibility of quantum biology should until now have seemed remote. Basic quantum phenomena (the wave-particle duality, the Heisenberg relations, the peculiar nature of quantum statistics, to name a few) until recently have been thought to occur only at the level of one or a few subatomic particles and to fade away in the presence of larger physical aggregates (like the ones studied in molecular biology). Equally, the need for an observer and/or his highly sophisticated instruments was an impediment. Biological phenomena seem to have no need of human observers. None seem to have been around when evolution began. In spite of such barriers, and others that might be mentioned, some

[1] I have struggled without success to find the locale of this statement in Whitehead's writings; and would be glad to be informed of it.

of the founders of quantum physics believed that in the end quantum physics would provide the real explanation for life. Among these were Erwin Schroedinger, Niels Bohr, and Walter Heitler (Anderson 2005, Begley 2006). It is now beginning to be conceivable that they were right.

The factors leading to this change in the conceptual landscape of biology are four:

1. The discovery of "mesoscopic" quantum effects: that is, quantum effects occurring at the level of large molecules and large collections of atoms.
2. Quantum coherence/decoherence theory, which obviates the need for a human observer (or the observer *qua* quantum instrumentation).
3. Quantum "nonlocality": the spooky action at a distance of which Einstein warned but which is now solidly ensconced in contemporary science.
4. Extremely accurate and detailed understanding of the molecular biology of the cell, cell membrane, subcellular parts (including DNA, RNA, amino acids, proteins…).

If anyone of these were lacking, quantum biology would not be possible.

1. Mesoscopic quantum effects

The emergence of quantum effects into the mesocosm comes as a surprise. The fundamental quantum phenomena now are seen to occur not only at the level of photons, protons, and neutrons but at the level of atoms and even large molecules.[2] The most striking example of which I am aware is Buckminster fullerene – the well-known "buckyballs". These are spherical molecules resembling soccer balls whose atomic structure incorporates either sixty or seventy carbon atoms: C_{60} or C_{70}. I cannot resist drawing two conclusions from such phenomena. First, if the quirky quantum effects occur in molecules of this size they obviously can occur in molecular-biological phenomena. Second, if such effects occur with molecules as

[2] Cf. Johnjoe McFadden's *Quantum Evolution* (158-159 and 225). All future references to this work will be cited in the text as QE.

massive as these, the impact of photons or electrons on the measured (observed) object cannot be called on to account for quantum effects. Photons and electrons, no matter how energetic (hence massive), cannot significantly affect molecules of such mass.

Among other mesoscopic quantum effects replicated in the laboratory are a situation in which six atoms "spin in opposite directions at the same time", and the "entanglement" of two groups of 100,000 atoms occurs some ten feet apart. Physicists W. Marshall, C. Simon, R. Penrose and D. Bouwmeister have designed an experiment which they hope will achieve a quantum superposition (a quantum wave state) nearly the size of a red blood cell (nine orders of magnitude more massive than any superposition observed to date) (Iredale 2004, Seife 2002). A growing and varied number of chemical reactions deviate from classical behavior because of quantum effects (McMahon 2003, Zuev et al. 2003).

2. No need for an observer

One of the more striking (and certainly most troublesome) aspects of classical quantum physics is its requirement for an "observer". The observer may be a human consciousness or an observing instrument or both. In any case its role is to observe the quantum state, collapsing the quantum wave (the superposition) into discrete simply-located particles. This peculiar dualistic structure has lead some (most notably Johann Von Neumann) to an idealistic philosophy of physics. Even if one does not go this far, however, the quantum observer is a puzzle. Biological evolution, as noted previously, somewhat antedates the appearance of human observers. Biological processes seem to be happening whether anyone is looking at them or not.

This latter suspicion is supported by recent coherence theory, which does not require an observer. According to this theory nature literally "measures" itself. That is, components of nature in the quantum state (quantum superpositions) are measured (the superpositions "collapse") when they come into contact with classical macroscopic objects. The interacting of the object in a quantum state with a classical, non-quantum state is termed decoherence. As we will see, Johnjoe McFadden's explanation of the emergence of life involves the consecutive interaction of coherence and decoherence. The main difficulty in making quantum computers is that the basic unit of quantum information, the "qubit", is a decoherent state which too easily collapses into a classical state (presumably by thermal factors). As Mae-Wan Ho insists, the notion of coherence is one of "cooperativity", not of atomism.[3] It is holistic.

3. Nonlocality

Einstein said two things about quantum physics that everyone seems to remember, the first that "God does not play dice with the universe", the second that if quantum physics is correct there must be "spooky action at a distance" in the universe. The great physicist today must be rolling in his retirement. Probabilities have not disappeared in quantum physics – indeed, they play a larger role there than ever. And the spooky action at a distance Einstein feared is beginning to receive experimental confirmation.

Action at a distance first appeared in Newton's gravitational law. For Newton, if at time t God created a physical object with a given mass (m), *instantaneously* the gravitational attraction exercised by this mass would be present throughout space, even at infinity (though at this distance, since gravitational attraction for Newton falls off according to the inverse square of distance, the attraction would be infinitesimal).

When quantum nonlocality enters the picture three factors enter with it: universal simultaneity, wave superposition, and action at a distance. But quantum action at a distance does not decrease with distance; distance is irrelevant. The example given above in which two groups of atoms ten feet apart are "entangled" is misleading. The same results would occur if the two groups of 100,000 atoms were ten light years apart, or even farther. Superposition, entanglement, and coherence provide possible explanations for biological phenomena at virtually all levels. As I will try to indicate later, they may change our notion of information as it functions in DNA/RNA/amino acids, of information transmission in the nervous sys-

---

[3] See Mae-Wan Ho's *The Rainbow and the Worm* (125). All future references to this work will be cited in the text as RW.

tem, and in the functioning of the entire organism. A kind of holism may emerge at all these levels, different from any yet considered.

4. Regnant complexity

Along with the three factors mentioned so far there is one more, equal in importance: complexity. I do not mean to use this term in the rather special way it is employed in discussing the "physics of chaos". The term complexity is used here in a more commonsense way. The notions of biological structure and function used when this writer was in high school and college were much simpler than they are today. In those far off days the cell membrane was conceived as being like a stretch of Saran Wrap, the cell was made up of something called protoplasm, and the genes were either mysterious entities hidden somewhere in the nucleus (or little dots on a blackboard explaining blue vs. brown eyes).

All this is now as quaint as the notion of phlogiston. The cell membrane is a differentially (and selectively) permeable membrane of extraordinarily complexity and dynamism. Protoplasm dissolves into a multitude of complex chemical compounds – a very varied multiplicity, in a complex synergy. And the genes, as everyone knows, step out of the shadows of our ignorance: DNA, RNA, amino acids, proteins (presumably perfectly understood) disport themselves in marvelous but very complicated ways.

That is, the cell, its nucleus, its non-nuclear parts if not perfectly known are known to us in great detail. This knowledge of detail is a triumph of scientific imagination, scientific research. But it opens up the possibility of connections previously unexpected: even inconceivable. I recommend that those interested in recently discovered connections read Eva Jablonka and Marion L. Lamb's *Evolution in Four Dimensions* for new discoveries in genetic material and genetic expression (2005, 462). *In nuce*: The new complexity provides innumerable possible junctures for the interaction of the new quantum physics with any number of newly discovered molecular features of the living organism.

To reiterate: the emergence of quantum effects into the mesocosm, the existence of a quantum physics not wedded to the observer, quantum nonlocality, and the discovery that the living organism fairly bristles with complexity at the molecular level: these factors jointly frame a new possibility, a quantum theory of the organism. But such a theory does not yet exist.

5. Three ways towards a theory

In what follows I would like to sketch the ideas of three physicists who are attempting to create such a theory. These are: Johnjoe McFadden, Mae-Wan Ho, and Peter Gariev. It goes without saying that none of these constitutes a complete theory.

McFadden's ideas are worked out in his *Quantum Evolution.* Of the three researchers to be examined, his ideas are the most straightforward. He begins with an explanation of that great enigma, the origin of life; then, to explain evolution, he proposes that the organism is a form of computer (a quantum computer).

His explanation of the origin of life takes advantage of the interaction of coherence and decoherence mentioned previously. He speculates that a small peptide molecule (in a protected place, e.g., a cavity in a rock) *via* the constructive force of coherence forms long peptide chains, alternating between linkage and *stasis*, as the chain interacts with matter in a non-quantum (classical) state. The process, McFadden argues, proceeds indefinitely.

> "[It would] have continued to elongate the quantum superposition of possible peptides until such a time when the system irreversibly collapsed into a classical state. The point at which this irreversible collapse would have taken place is easy to predict: it would have been when the peptide learned to self-replicate" (QE 227).

Once self-replication was established, Darwinian natural selection would take over, pruning mutant molecules, creating "superior" next generations.

But the fundamental cause of evolution would be neither point mutation nor natural selection, but the cell's and/or the organism's capacity to perform as a quantum computer: a very particular kind of quantum computer. Orthodox accounts of the function of DNA (bearer of biological information) leave out a possibility: namely, that internal quantum measurement confers on the living cell "an ability to influence its particle dynamics in a way unique to life" (QE 255). It may even be able to create – in the terms of geneticist John Cairns – "directed" mutations (QE 263-264).

Most of those working to constitute a quantum biology accept the metaphor of the quantum computer. The exception would be Mae-Wan Ho, who nowhere raises the question of quantum computation in the organism. Paul Davies, in a recent editorial in *Nature*, argues that the organism is clearly a computer capable of "measuring" its internal states, and observes:

> "Molecular biologists are content with ball-and-stick models based on classical concepts. But so long as they cling to that, the origin of life will remain mysterious" (2005, 819).

But a quantum computer differs from our present digital computers not only in being more powerful: the answers it gives are less definite, more open to interpretation (within limits). The organism does not become a p.c.

Mae-Wan Ho's concept of the quantum organism differs from McFadden's somewhat in theory, very much in approach. Ho's work and emphasis are focused on laboratory experiment. These lead her first of all to suggest a new kind of thermodynamics, in addition to the classical thermodynamics of Ludwig Boltzmann or the newer nonlinear thermodynamics of Ilya Prigogine and his colleagues. This third thermodynamics rests on the extraordinary efficiency of many biochemical processes, notably photosynthesis and muscle contraction, which, literally, produce no entropy at all. Hence, relieved of the need to pay debts in entropy, the organism is easily able to store energy without using it up. Hence, she argues, "organisms are anti-entropic as long as they are alive" (RW 73).

A recent experiment at the University of California, Berkeley makes an important step towards demonstrating the dependence of photosynthesis on quantum wave phenomena (Engel 2007). This result both supports Ho's theory that quantum physics plays an important role in biology and her contention that there is a new, third sort of thermodynamics intimately connected with quantum physics. Some thinkers (Joseph Early, this volume) appear to believe that Prigogine's nonlinear thermodynamics by itself provides the basis for the biology of the future without help from other parts of physics. The ultimate relations between thermodynamics and quantum physics, however, are now simply not known. It would be wrong to exclude quantum physics from the picture at the present time.

Two other factors are fundamental to Ho's biology. One is her belief that the organism is a liquid crystal:

> "[…] liquid crystals have an orientational order, in that the molecules are aligned in some common direction(s), rather like a crystal. But unlike solid crystals, liquid crystals are flexible, malleable, and responsive" (RW 173).

All major components of the living system DNA, RNA, muscle proteins, collagens, connective tissues are for Ho liquid crystals. Both her third type of thermodynamics – the "thermodynamics of organized complexity" (RW 79, 85) – and her conception of the organism as a liquid crystal have a common virtue. Both make possible communication within the organism by means of very weak signals.

Hence her belief that quantum coherence can account for the unity and viability (the wholeness and the capacity to sustain itself) of the living organism. That is, given the presence of a third level of remarkably efficient energy use and storage and the easily shaped responsive liquid crystal state, it becomes possible to understand how a quantum coherent field can affect a living thing:

> "[…] the living system is one coherent 'photon' field bound to living matter. This photon field is maintained far from thermodynamic equilibrium and is coherent simultaneously in a whole range of frequencies that are nonetheless coupled together" (RW 152).

That is, each of the different components of the organism has its own rhythm. These rhythms are made synchronous by a quantum-coherent field.

It is important to point out a contrast between Ho's position and that of some other champions of quantum biology. Roger Penrose and Stuart Hameroff propose that complexes of microtubules in neurons provide an environment sufficiently protected to make quantum entanglement and superposition possible, and thus explain the possibility of indeterminism in the organism's behavior (Joseph Early, this volume). Ho has no need for this hypothesis. For her quantum fields by themselves can account for free behavior.

I have not been able to put together enough information on Peter Gariaev to give a full report on his research. He has published a book in Russian, but my Russian is not equal to the task of translation. The materials available to me consist of articles taken from the Internet. Gariaev terms himself the father of wave-genetics (2005), a theory that he formulates as follows:

1. The evolution of biosystems has created "texts", similar to natural context dependent texts in human languages, shaping the text of these speech-like patterns.
2. The chromosome apparatus acts simultaneously both as a source and receiver of the genetic texts, respectively decoding and encoding them.
3. The chromosome continuum of multicellular organisms is analogous to a static-dynamical multiplex time-space holographic grating, which comprises the space-time of an organism in a convoluted form.

The DNA molecule, on Garaiev's terms, is thus able to form three-dimensional (and holographic) images both of the biostructures which make up the organism and of the organism as a whole. It thus resembles a homunculus far more than it does the successive dispenser of a triple codon language. The action of DNA is for Garaiev typically nonlocal: the quantum nonlocality encountered several times previously in this essay. It can appear 1) as with Ho, in the organism as a whole 2) at the molecular level 3) at the cellular-nuclear level 4) at the cellular level, and 5) at the "chromosome-holographic" level. Hence (again as with Ho) "bioinformatics events can be instantaneously coordinated". They take place "here" and "there" simultaneously. The intercellular diffusion of signal substances is too slow to account for how highly complex biosystems work in real time (Gariaev et al. 2002).

6. Some philosophical implications

I must call a halt to this essay. To go into detail would require more space than is possible here. I would like to conclude by suggesting some present successes on quantum biology, as well as some of its challenging philosophical implications.

The theories of the three figures sketched above are speculative. That does not mean that they are wrong. The four sets of circumstances described at the beginning of this essay frame a real possibility for quantum biology. The three authors attempt to construct theories within this frame. The laboratory successes of Ho and Garaiev also deserve mention. I will mention here only one. Garaiev and his associates, using orthogonal "biophotons" refracted from the genetic material of the lowly carrot produces both structural abnormalities on the surface of the carrot and markedly greater growth rates. A shower of photons *not* refracted from the genetic material of the carrot produces no results at all.

In several respects quantum biology is markedly congenial with a Whitehead philosophy of nature. The quantum universe, like the Whiteheadian, is a world of actual occasions which through their dynamic interrelations produced new complex entities. Nowhere in such a world are there fundamental hard, solid particles moved by impact and externally related. In such a world there are no organisms which can in any simple way be reduced to their parts. Holism (a unique sort of holism) reigns.

But it would be wrong simply to identify Whiteheads vision of nature with the present (admittedly incomplete) quantum biology. Some reworking of the Whiteheadian metaphysics seems to be required. For one, while actual occasions for Whitehead were limited to subatomic quantum events, the newer extensions of quantum physics give us actual occasions of far greater extent, both spatially and temporally. Just how great these extents might be is currently unknown: *there appears to be no upper limit.*

This, I think, is not a fundamental problem for Whitehead. Two factors basic to quantum biology, however, might require deeper consideration. I refer both to the probability wave and to nonlocality. I do not find the latter in Whitehead's philosophy (which is based on the pre 1924-1927 formulations of quantum mechanics). Equally, there is in Whitehead no "action at a distance". Transfer of characteristics in his universe occurs through prehensions, and these are local in nature. To deal with such factors might force some widening and rethinking of Whiteheadian categories.

REFERENCES

Anderson, P. (2005). "Thinking Big". In: *Nature*, 437, September 9, p. 625.

Begley, S. (2006). "The Magical Behavior of Subatomic Particles Moves into the Real World". In: *Wall Street Journal*, 247, n.4, January 6, A9.

Davies, P. (2005). "A Quantum Recipe for Life". In: *Nature*, 437, n.7060, October 6, p.819.

Engel, G. et al. (2007). "Evidence for Wavelike Energy Transfer through Quantum Coherence". In: *Nature*, 446, n.7137, April 12, pp. 782-786.

Gariaev, P. (2005). "Open Letter from Peter Gariaev, the Father of Wave-genetics". http://www.fractal.org.Life-Science-Technology/Peter-Gariaev.htm
This item is republished from the September, 2005 issue of *DNA Monthly*.

Gariaev, P. et al. (2002). "The DNA-wave Biocomputer" http://www.rialian.com/rnboyd/dna-wave.doc

Gunter, P. A. Y. (2006). "Darwinism: Six Scientific Alternatives". In: *The Pluralist*, 1, n.1, pp. 13-30.

Ho, M. W. (1998). *The Rainbow and the Worm*. 2nd Ed. New Jersey: World Scientific.

Iredale, M. (2004). "Is it Time Schrödinger's Cat was let out of the Bag?" In: *The Philosopher's Magazine*, 1st Quarter, p.16.

Jablonka, E.; Lamb, M. (2005). *Evolution in four Dimensions: Genetic, Epigenetic, Behavioral, and Symbolic Variation in the History of Life*. Cambridge, MA: The MIT Press.

McFadden, J. (2000). *Quantum Evolution*. New York: W.W. Norton.

McMahon, R. (2003). "Chemical Reactions Involving Quantum Tunneling". In: *Science*, 299, February 7, pp. 833-834.

Seife, C. (2002). "Quantum Experiment Asks 'How Big is Big?" In: *Science*, 298, October 11, pp. 342-343.

Zuev, P. et al. (2003). "Carbon Tunneling from a Single Quantum State". In: *Science*, 299, February 7, pp. 867-870.

The Effect of Mind upon Brain

HENRY P. STAPP

A physics-based understanding of how our conscious thoughts can affect our physically described brains is presented. This understanding depends on the shift from the mechanical conception of nature that prevailed in science from the time of Isaac Newton until the dawn of the twentieth century to the psychophysical conception that emerged from the findings of Planck, Bohr, Heisenberg, and von Neumann. This shift converted the role of our conscious thoughts from that of passive observers of a causally closed physically described universe to that of active participants in an essentially psychophysical understanding of nature.

1. The basic question and why it is important

Science's conception of the physical world changed radically during the twentieth century. At the end of the nineteenth century most scientists still viewed the physical universe as essentially a giant machine. This mechanical view emerged from the seventeenth century work of Isaac Newton, who built his conception of nature on the ideas of René Descartes. According to Descartes, the universe is composed of two kinds of elements, the first consisting of elements each of which occupies at each instant of time a definite region in space, and the second consisting of elements that include our human thoughts. Descartes allowed these two parts of nature to interact causally within our brains, but Newton specified that the motions of the elements of the first kind are completely determined by laws of motion that refer exclusively to elements of this kind. In Newtonian-type physics these elements of the first kind are, moreover, essentially mechanical, in the sense that they have been stripped of the experiential qualities that characterize our thoughts, such as the conscious awareness of feelings, and the capacity to grasp meanings. Consequently there is in Newtonian-type physics no possibility for any causally effective role for any aspects of our being that are essentially different in kind from the mechanical elements that occur in that type of physics. Descartes' views accommodated our intuitive feeling that our conscious thoughts can influence our bodily movements – my conscious intent to raise my arm *seems* to cause my arm to rise – whereas Newton's physics leads to the contrary conclusion that, in spite of *what may seem to be the case*, the notion that human thoughts, per se, can affect human actions is a deeply misleading illusion.

Newton's ideas led eventually to the classical physics of the late nineteenth century. Its main premises are these:

1. There exists a material universe that develops over the course of time by means of interactions of its tiny mechanical parts with neighboring tiny mechanical parts.
2. These interactions are governed by mathematical laws.
3. These laws entail that the mechanically described future is completely determined by the mechanically described past, with no reference to human thoughts, choices, or efforts.

This conclusion is called: *The principle of the causal closure of the physical.*

This "principle" seemed at one time so secure, and so central to the scientific enterprise, that some scientists came to view science as not essentially an open-minded empirically based inquiry into the structure of nature, but as also an ideology: i.e., as the idea that scientists must be tenacious defenders of the dogma that we human beings are essentially material/mechanical systems governed exclusively by matter-based laws, and hence that our conscious thoughts and intentions can have no actual effects upon our physical actions: that the universe is, with respect to its basic causal structure, completely mindless.

Within that earlier pre-twentieth-century science-based conception of nature the existence of our streams of conscious thoughts constituted a major embarrassment. The occurrence of things having the defining characteristics of our conscious thoughts was in no way entailed by the properties of the physical world that the physicists had postulated. Our thoughts, ideas, and feelings could be imagined to be produced – in some unexplained way – by the complex activities of our brains. But there was no logical basis in the classical physicists' conception of nature for understanding or explaining the emergence of human experience. Although philosophers wove endless tapestries of words in an effort to relate the physically described aspects of the world to the experientially felt aspects of our lives, the efforts of those thinkers were invariably judged inadequate by their critically minded colleagues. Insofar as our brains were understood exclusively in terms of the concepts of classical physics a causal gap persisted: a conceptual chasm remained between, for example, a painful feeling and the associated activities – no matter how complex and novel – of the associated physically described body and brain.

During the twentieth century this classical-physics-based conception of the world was found to be logically incompatible with a growing accumulation of empirical data. Eventually, the classical mechanistic description of the physical aspects of nature was replaced by the profoundly different quantum mechanical description.

The orthodox formulation of quantum mechanics, which is the form used in all practical applications, was created by Heisenberg, Bohr, Pauli, and Born during the 1920's. Shortly thereafter it was cast into a more rigorous logical and mathematical form by the logician and mathematician John von Neumann.

Quantum mechanics differs from classical mechanics in deep mathematical ways. In order to tie the new mathematical structure to empirical data in a practically useful way the founders of quantum mechanics instituted a profound break with one of the basic principles of classical physics: they inserted the conscious experiences of human beings into the dynamical workings of the theory. Human beings were allowed, and indeed required, to act both as causally efficacious agents, and also as causally efficacious observers. In particular, orthodox quantum mechanics requires every conscious observation to be preceded, logically, by an action that specifies a "Yes-or-No" question, which a feedback "observation" will then answer either by a "Yes" or by a "No". Both of the two actions, the query and the feedback, are causally efficacious: they alter in different non trivial ways the physically described state of the universe.

Each of these two actions is described in two different ways. Each action is described first in the psychological language that we use to communicate to each other, and to ourselves, the structure of our experiences.

And it is described also in the mathematical language of quantum physics. Each psychologically described event becomes thereby linked, *within the theory*, to the quantum mathematical description of the physical world. This dualistic – psychophysical – description was needed to link the quantum theoretical description of the mathematically controlled evolution of the physically described world to the structure of the communicable human experiences that constitute the empirical basis of science.

In this quantum mechanical description the unfolding of the universe is no longer governed by the physically described aspects of nature alone. Neither of the two actions, neither the query nor the feedback, is determined *within the orthodox theory* by prior physically described sufficient conditions. Within orthodox quantum mechanics, our causally efficacious conscious intentional efforts remain "free" of any specified physical coercion. Yet these conscious efforts do have, according to quantum mechanics, important physically describable effects. Quantum mechanics thereby rescinds the materialist dogma: it *falsifies* "the principle of the causal closure of the physical"! But in spite of this loss of its scientific support, the classical materialist ideology, including the precept of causal closure of the physical, continues to infect the thinking of many contemporary scientists and philosophers: the materialist ideology has survived the death of its body of scientific support

But why are these seemingly arcane matters important? Why, in the context of the pressing contemporary concerns of the human race, is attention to these scientific issues worthwhile? It is profoundly worthwhile because science's pronouncements on the nature of our own human beingness, and on the character of the connection of our conscious intentional efforts to the unfolding physical reality, underlie much of the contemporary discourse on urgent societal issues.

The classical-physics-based conception of human beings has had a tremendously pernicious impact on our cultural heritage, because it paints us, on the one hand, as mechanical automata, whose conscious intentional efforts can have no causal effects whatever on the physically described aspects of nature, and, on the other hand, as mechanical consequences of a dog-eat-dog competition for survival. The consequence of the first effect is to discourage conscious effort to improve the human condition as pointless and irrational; and the consequence of the second effect is to justify unrestrained self-aggrandizement at the expense of the essential well-being of others. The materialist dogma undermines the foundations of moral and ethical philosophy.

Our beliefs about our relationship to the world around us underlie our values. And our values determine the sort of world we strive to create. The main social problems we face today stem primarily from the fact that different approaches to this basic question of our place in nature lead to different conclusions, and hence to conflicting values, and thence to conflicting actions. Because of the stature of science in contemporary culture it is vital to answer as accurately as possible the question: What does contemporary basic physics say about the nature of the connection of our conscious thoughts to the physically described aspects of nature?

2. From the classical to the quantum mechanical conception of the role of human beings in the unfolding of reality

Quantum mechanics rests upon a mathematical foundation provided by classical mechanics. The latter rests upon the idea of "particles" and "fields". A particle is supposed to have, at each instant of time, a definite position and a definite velocity in three-dimensional space. A field is supposed to have, at each instant of time, and at each location in three-dimensional space, a definite "value", specified by a real number. The field variables are connected to the particle variables in way that determines the forces upon – and hence acceleration of – each particle, due to the presence and the motions of the other particles. These positions and numbers, together with a few constants that determine such things as the masses of the particles and the strengths of various forces, constitute the *physical aspects of nature.*

Newton conjectured the existence of repulsive forces that prevent particles from coming too close to each other. This condition combined with his other laws appears to entail "the causal closure of the physical": the description of the physical aspects of nature over some short interval of time, combined with the physical laws, completely determines all physical aspects of nature for all times.

This closure feature allows the evolving state of the universe to be pictured as a block physical universe; namely by a collection of particle trajectories – conceived as infinitely thin "wires" – running through the space-time, in the direction of increasing time, and in a way that is uniquely determined for all times by the form of this physical structure over any short interval of time. (The "fields" should also be represented, but their pictorial image is slightly more complicated.) In Newtonian-type mechanics no representation of experience or knowledge need be considered. That is why this feature of Newtonian-type physics is called "the causal closure of the physical".

The transition from classical – Newtonian-type – mechanics to quantum mechanics brings human knowledge and experience importantly into the theoretical framework. The reason, basically, is this: the way the mathematical/physical description enters into practical applications is closely analogous to the way that the mathematical/physical description enters into classical *statistical* mechanics. But classical *statistical* mechanics is, in its practical applications, closely tied to human knowledge: A sudden change in "our knowledge" causes a sudden change in the mathematical/physical representation of our knowledge, which produces in turn a sudden change in the knowledge-based predictions of the theory.

Classical statistical mechanics accommodates in a completely understandable way our uncertainties about the actual positions and velocities of the physical particles. An analogous feature of quantum mechanics is the "Heisenberg Uncertainty Principle". The effect of this principle is, essentially, to convert each "wire" of the block universe picture into a smear of weighted possibilities. More precisely, for a many-particle universe, the effect of the Heisenberg uncertainty principle is to replace the one single actual many-particle universe of classical Newtonian-type physics by the collection of all such (weighted) possibilities compatible with the present state of "our knowledge". This smearing out of the set of possibilities can be large at the level of the atomic particles. Then, because of the sensitive dependence of macroscopic degrees of freedom upon microscopic initial conditions, the so-called butterfly effect, the smearing out at the *macroscopic* level tends to increase with the passage of time. Yet at certain moments we gain, via our (sense) experiences, new knowledge. Just as in classical statistical mechanics, this new knowledge will usually exclude some of the possibilities that were mathematically generated from an earlier state by the mechanical equations of motion This sudden gain in knowledge will be represented by a sudden "collapse" of the mathematical representation of the state of our knowledge to a "reduced" form: to a "reduction" of the size of the "wave packet".

This reduction of the size of the physically described wave packet is tied, just as it is in classical statistical mechanics, to the increase in our knowledge. For example, if we originally know only that a particle is in a certain box, and then learn, from some observation that we make, that the particle is definitely not in the right-hand half of this box, then the region over which the weighting factor is non-zero is "reduced" to the left-hand half of the box. This reduction, combined with our knowledge of how the physical state evolves during the intervals between our observations, will alter our expectations about our future experiences.

There is nothing at all mysterious about such sudden "collapses" in classical statistical mechanics. The "collapses" that occurs in quantum mechanics are, *at the level of actual scientific practice*, quite analogous to it: in both cases the mathematical representation of "our knowledge" changes abruptly when "our knowledge" changes abruptly. It is this close correspondence – *at the level of actual scientific practice* – of quantum mechanics to classical statistical mechanics that allows scientists to use quantum mechanics in a rationally coherent that they can understand intuitively.

But there is a conceptual problem: in quantum mechanics the different "classically conceived possibilities" can interfere with each other in ways that they cannot do in classical statistical mechanics. According to the precepts of classical statistical mechanics, all but one of the various weighted possibilities exist only as figments of our imagination, and the one real situation cannot be affected by possibilities that exist merely as imagined possibilities – which we have found useful to contemplate because of our lack of knowledge about which of the possibilities allowed by the physical theory is the unique one that really exists. Classical statistical mechanics demands, accordingly, that the various possibilities evolve independently of each other, in the way physically specified by the underlying classical mechanics. This postulated causal structure effectively blocks any causal effects of our human thoughts *per se* – which are characterized by conscious awareness, feelings, and understandings of meanings – upon the dynamical machinations of the physical aspects of nature. According to classical mechanics, these physical aspects are completely controlled exclusively by these physical aspects themselves, which have been completely stripped by the precepts of classical mechanics of those features that characterize our thoughts. From the standpoint of classical physics, our strong intuitive feeling that our thoughts themselves can affect the physically described world must be deemed a misleading illusion.

The claim of classical physics that there is this causal disconnect between the mental and physical aspects of nature is so peculiar and unnatural as to render nearly irresistible the idea that this physical theory must be an approximation eventually to be superseded. Although some thinkers continue to cling steadfastly to the precepts of nineteenth century classical physics, which automatically enforce the very approximation that needs to be undone, those precepts were shown already during the first part twentieth century to be incompatible with empirical findings, and to be, moreover, indeed an approximation to a more adequate physics – quantum physics – that does in fact feature essential dynamical links between these two aspects of nature.

The resolution of the conceptual problem mentioned above is to interpret the quantum mechanically described state of the universe as a representation not of *imagined physical possibilities,* but rather of "*potentialities for future psychophysical events*": i.e., as a representation of *objective tendencies*, created by past psychophysical events, for the occurrence of future psychophysical events. This interpretation is essentially implicit in orthodox quantum mechanics, and was nicely described by Heisenberg (1958 a, b). It transforms "our knowledge" from a realm of realities deemed unable to causally affect the unfolding of material/physical/mechanical aspects of reality to an essential causal input into the unfolding of an integrated psychophysical reality.

Of course, science has always been about "our knowledge" in a certain ultimate way. It is about what we can know, and how we can use what we know to affect what we will probably experience under the various alternative possible courses of action between which we are seemingly free to choose. However, the effect of Newton's monumental work was effectively to banish the human psyche from any causal role in its own future. Consequently, quantum physics is, perhaps more importantly than anything else, a liberator of the human mind from the 200 years of bondage imposed, within science-based philosophy, by the huge achievements of Isaac Newton.

The main interpretive idea of orthodox quantum mechanics is that each acquisition of knowledge occurs discretely in conjunction with “a collapse of the quantum state” to a new form that incorporates the effect of adding the conditions logically imposed by the increase in knowledge. This change forges a tight logical link between “an experientially recognized change in a state of knowledge” and a corresponding “mathematically represented change in the physical state of the universe”. The new psychophysical state represents both a new state of knowledge and also a related new set of statistically weighted potentialities for future psychophysical events.

2.1. Von Neumann’s shift of the Heisenberg cut

The original “Copenhagen” interpretation of quantum mechanics separated the physically described world into two parts: (1), the system being probed, which is considered to consist of atomic constituents described in the mathematical language of quantum mechanics; and (2), the rest of universe, which is treated as the “observer”, whose experiences pertaining to the observed macroscopic components of the world are described in the language of classical physics. This observing portion is supposed to include both the human observers and their macroscopic measuring devices, conceived and treated in the way that classical physics conceives and treats macroscopic objects. However, the macroscopic devices and the human observers are themselves composed of atomic constituents. Hence it is not clear where one should place the boundary between the part of the world that is described in terms of the quantum theoretical mathematics and the part that is described in terms of human experiences that can be expressed in terms of the language of classical (Newtonian-type) physics. There is no basic principle beyond practical utility that precisely specifies the placement of this “Heisenberg cut”.

This question was studied by non Neumann (1932/1955), who showed, at least within the framework of the idealized solvable model that he was examining, that the predictions of quantum theory are invariant under a shift of a macroscopic device from the classically described side “above” of the cut to the atomistically described side “below” the cut. By a series of shifts von Neumann includes more and more of the world in the part "below" the cut, until at last the entire world that can be conceived to be built of atomic particles, including the entire brains of all the observers, are described in terms of the quantum mathematics. Thus Von Neumann examined the problem of where to place the cut by considering an idealized situation in which there is a sequence of measuring devices, each probing the output of the device that precedes it in the sequence, and by then following the causal chain first into the retina of the observer, and then into the optic nerves, and then ever deeper into the brain until at last the entire brain of the observer is treated quantum mechanically, along with the rest of the physical universe.

Yet quantum mechanics was formulated from the outset in terms of the relationship between the two different descriptions, the physical and the psychological. This dualistic structure continues to be maintained at each shift of the boundary, even when the entire physically described world is treated quantum mechanically. Indeed, what needs to be preserved at each step is essentially the predicted relationships between human experiences, which are described in psychological terms. The final part of the brain that remains just above the cut before the final shift is described in psychological and classical terms just before the final shift, but in atomic quantum-mechanical terms after that final shift. Von Neumann gives the name "abstract ego" to the carrier of the logically needed psychologically described aspects that must still remain even after all the brains of all the agents are described purely in physical terms. The theory at that stage describes the entire physical world quantum mechanically, with each psycho-physical event now specifying a particular action, associated with a definite spatial region and a certain definite time, that connects the mind and the brain of a conscious agent.

The nature of this psychophysical connection will be described presently. But the upshot is that this connection allows, and indeed seems to require, the psychological events that populate our streams of conscious experience to play a dynamically active role in the determination of the temporal development of the physically described properties of our brains. This psychophysical connection, which is the focal point of this article, will now be discussed in more detail.

2.2. The basic conflict between classical and quantum physics

Classical mechanics assumed that the ideas that work well for large objects, such as planets, moons, and falling apples, will continue to work all the way down to the level of the atoms and molecules. According to this classical notion, each particle, such as an electron, has a well defined trajectory in space-time. This idea is illustrated in figure 1.

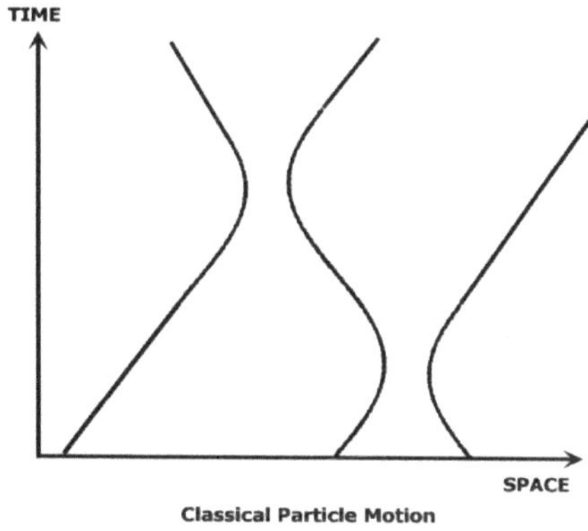

Fig. 1: *Classical Physics.* This diagram shows a possible evolution in time of a system consisting of three classically conceived electrons. Each particle has a well defined trajectory in space-time, and each particle repels the others increasingly as their trajectories come closer together.

The laws of motion of classical physics ensure that the trajectories of all the particle (and fields) in the universe at times earlier than some fixed time t fix the trajectories of all particles for all future times.

A principal change introduced by quantum theory is the "quantum uncertainty principle". This principle asserts that each particle must be represented, NOT by one single well defined trajectory, but by a cloud of possible trajectories, as is shown in figure 2.

The effect of these uncertainties, if left unchecked, would be disastrous. The uncertainties at the atomic level tend to bubble up, irrepressibly, to macroscopic levels. If the uncertainties originating at the micro-level were left unchecked from the time of the "big bang", the macroscopic world would be by now a giant cloud encompassing all possible worlds, in stark contrast to the essentially single macroscopic world that we actually observe. For example, if the uncertainties were left unchecked then the moon would be spread out over much of the night sky; And each person's brain would correspond to a mixture of all of the many alternative possible streams of consciousness that the person could in principle be having, instead of corresponding to the essentially single stream of consciousness that each of us actually experiences.

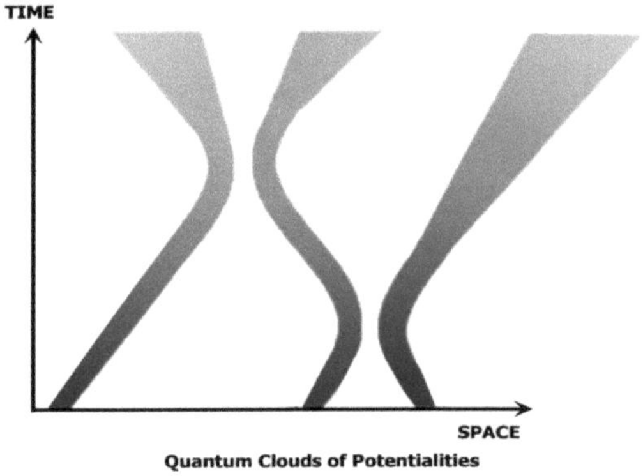

Fig. 2: *The quantum cloud.* This diagram illustrates the usual effect of "quantizing" the classical system shown in figure 1. Each trajectory line of the classical theory is broadened out into a cloud of alternative possible trajectories, as a consequence of the uncertainty principle.

To deal with this difficulty the founders of quantum theory were forced to draw a clean conceptual distinction between the two aspects of scientific practice, the empirical and the theoretical, and to introduce a special process to account for their connection. The empirical component describes our experiences pertaining to what we human beings do, and to the experiential feedbacks that we then receive. The theoretical component describes the "particles and fields". The process that connects these two aspects of the scientific description of the world is called the process of measurement or observation.

This measurement/observation part of the quantum mechanical approach erects a firewall that protects the empirical/experiential realm from an unfettered intrusion of quantum uncertainties from the theoretical realm.

2.3. The firewall that protects the empirical realm from the quantum uncertainties that pervade the physical realm

But how are the quantum uncertainties held in check? The theory of the process of measurement was put into rigorous form by John von Neumann, building on ideas of Werner Heisenberg.

The theory demands that each experience occurs in conjunction with an associated *intervention* into the Schroedinger-equation-controlled evolution of the physical state. This intervention is called "process1" by von Neumann. This physical action is associated with a psychological element, which is a specification of a particular question that can be answered empirically by either a "Yes" or by a "No". (Multiple choice questions can be reduced to sequences of Yes-or-No questions.)

The temporal evolution of the full reality is punctuated with these process-1-related events. Each such event has two related aspects, one in the empirical domain of "our knowledge", and the other in the domain of the mathematical description. On the empirical side the action specifies a certain possible "increment in knowledge": an experientially recognizable "Yes" response to the question. This "Yes" answer is linked on the mathematical side, to a reduction of the prior quantum mechanical state to that part of itself that is consistent with the increase in knowledge corresponding to the answer "Yes". This reduction is analogous to the similar reduction that occurs in classical statistical mechanics, as has already been explained.

If nature fails to deliver the answer "Yes", then the prior physical state becomes reduced to the part of itself that is unambiguously associated with the answer "No". This action process is represented in figure 3.
Von Neumann calls *the physical part* of this action by the name *Process 1*.
Two important facts about process 1 are:

1. The process 1 actions enter importantly into the dynamics.
2. Quantum mechanics does not identify any logically sufficient physically described cause for this action!

Consequently, the "principle of the causal closure of the physical" is not entailed by the rules and content of orthodox quantum mechanics!

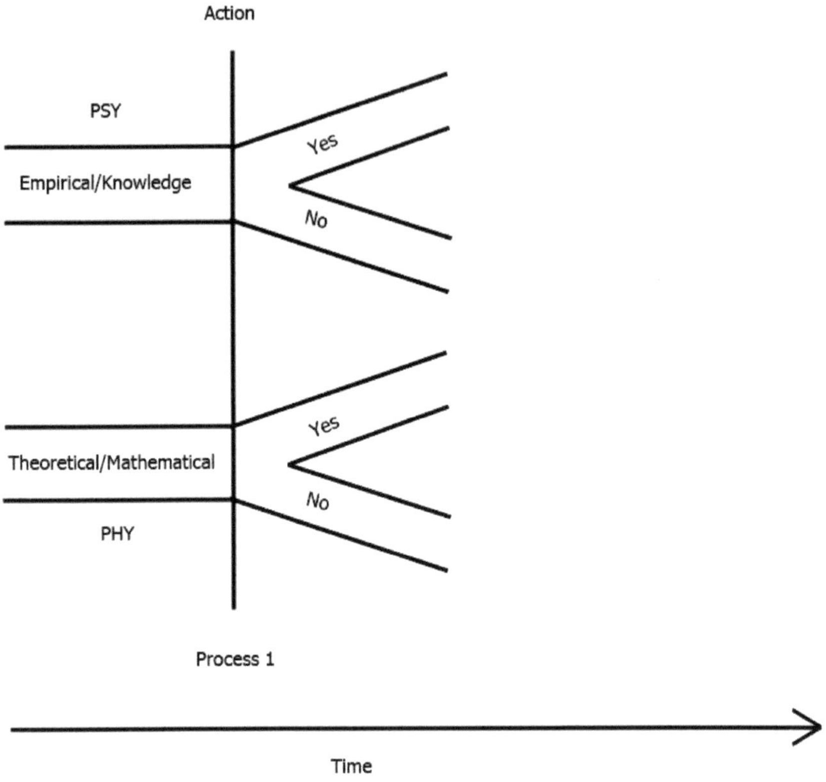

Fig. 3: *The action associated with Process 1.* Each empirical finding is associated with a process 1 quantum reduction event that has both an experiential part that represents a separation of the agent's stream of consciousness into two particular alternative possible paths, labeled "Yes" and "No". The "Yes" answer identifies an experimentally/empirically defined "Yes" feedback to an agent-posed query. There is a corresponding separation of the physical state into a "Yes" part and a "No" part, with the "Yes" part specifying the part of the prior physical state that is compatible with the agents having a "Yes" experience.

Von Neumann gives the name *Process 2* to the process that causes the quantum state to evolve according to the Schroedinger equation, which is the quantum analog to the classical deterministic equations of motion.

The measurement process has a second part: the Yes-or-No feedback from the associated action. This second stage is called *Process 3*, and is pictured in figure 4.

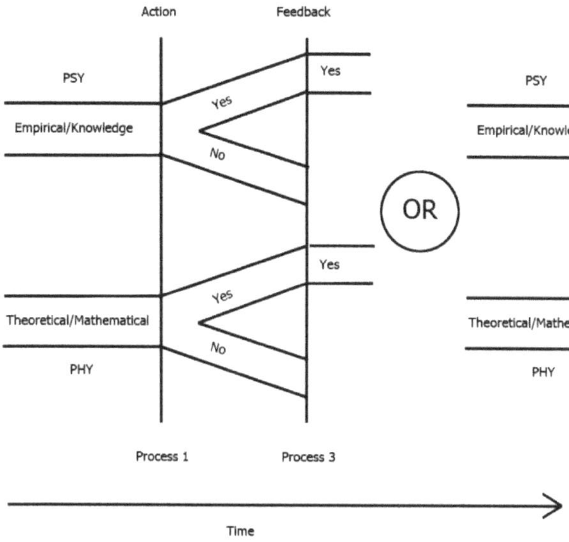

Fig. 4: *Process 3*. Nature delivers a feedback "Yes" or "No" to the agent's query.

According to quantum mechanics, the feedbacks conform to statistical conditions that are specified by the theory. The choice of the feedback (or outcome) is what Dirac called: "a choice on the part of nature". According to quantum mechanics, this choice of outcome is statistical, and, unlike process 1, it lies outside the hands of human beings.

The choice of the process 1 probing action is what Heisenberg called "a choice on the part of the 'observer' constructing the measuring instruments and reading their recording" (Bohr 1958, 51). As regards this choice Bohr says:

> "The freedom of experimentation […] corresponds to the free choice of experimental arrangement for which the quantum mathematical formalism offers the appropriate latitude" (Bohr 1958, 73).

These remarks by the founders of quantum mechanics emphasize that the physically described aspects are, within contemporary physics, no longer dynamically complete, but have, in particular, a process-1-related dynamical gap that, at least in practical applications, is considered to be influenced by our consciously experienced intentions.

The process of measurement creates a firewall that blocks the unfettered diffusion of the quantum uncertainties into the empirical (or experiential) realm. *It is the choice of a process 1 action, which is not controlled by any known physical process, statistical or otherwise, but which appears to be influenced by understandings and conscious intentions, that, in conjunction with a stochastic process 3 choice of feedback on the part of nature, controls which potentialities pass through the firewall, and into the realm of our actual experiences!*

I call the process, whatever it is, that chooses the form of process 1, and the time *t* at which the process 1 action occurs, by the name *Process Zero*. The fact that process zero is not determined by the physical laws of contemporary quantum theory constitutes a "causal gap" in that theory: it entails an apparent breakdown of the "principle of the causal closure of the physical". The orthodox interpretation of the theory is designed for practical application, and it fills this causal gap at the physical level by allowing an *intervention* from the mental realm to create a filter/firewall that permits only certain definite thoughts to emerge from the uncertain indefinite prior physically described universe.

3. Template for action

Any intentional physical action, such as raising one's arm, requires sending a temporally correlated sequence of neural signals to the muscles. So it is plausible that there is, in association with each intentional action, a corresponding spatio-temporal pattern of neural or brain activity that if sustained for a sufficient period of time will tend to cause that action to occur. I call this spatio-temporal pattern of brain activity a *template for action*. The action of the process 1 associated with this intentional action should preserve this template for action, and eradicate the possible patterns of brain activity that are incompatible with it.

3.1. Process 1 and the conversion of an intentional thought to a bodily action

An experimenter's action of setting up a particular experiment is an action directed at the goal of receiving an intended feedback. It is represented in quantum theory by a psycho-physical event. The psychological aspect is the felt intention to receive the intended feedback. The "Yes" part of the physical aspect of the associated process 1 action reduces the state of the brain to the part compatible with the template for the intended action.

3.2. The quantum Zeno effect

It is a consequence of quantum dynamics that sufficiently rapid repetitions of the same process 1 action can, by virtue of the so-called quantum Zeno effect, cause the template for an intended action to be held in place, in the face of strong disrupting physical forces, for much longer than would otherwise be the case. Such an extended holding-in-place of this template for action will tend to make the intended action occur. *Thus influencing the repetition rate of a sequence of process 1 actions can influence the bodily actions of the agent.*

The repetition rate of a sequence of process 1 actions in a human brain is not controlled by the known quantum physical laws. Thus we are invited to consider the possibility that these repetition rates can be influenced by mental realities, and in particular by the psychological intensity of the conscious intention to perform the physical action associated with a template for action. Postulating such an influence creates the possibility of accounting causally, within contemporary physics, for the apparent connection of conscious intention to bodily action; a perceived connection upon which we base our entire lives. Classical physics, by restricting causal connections to a causally closed physical domain, with no conceptual foothold for any logical link to consciousness, provides no analogous possibility for rationally understanding a real effect of our conscious intentions themselves upon our physical actions.

The process zero, whatever it is, that determines the form and the timings of the process 1 actions is *not* the quantum mechanical process, process 2 (the Schroedinger equation controlled process) that governs the time evolution of the quantum state of the system. That quantum state, according to orthodox ideas – particularly those of Heisenberg – specifies only the potentialities/probabilities for the actual events, but neither the form nor the timing of the *actual* events themselves. The Schroedinger equation, which is a quantum mechanical analog of the equations of motion of classical mechanics, makes no reference at all to idea-like realities such as intentions or mental concepts. But this same limitation does not apply to process zero. *Insofar as process zero allows the repetition rate for a sequence of similar process 1 actions to be influenced by conscious intentions, quantum mechanics provides a fundamental-physics-based way for our conscious intentions to bring the physical correlates of our conscious intentions into the physically described universe.*

Quantum mechanics leads, therefore, to a radical revision of our conception of ourselves. Whereas classical physics reduced man to an essentially mindless machine, quantum mechanics allows a person's conscious intentions to influence his physical actions.

Once this power of our consciously felt intents to causally influence the experienced feedbacks is introduced into the dynamics, the agents become enabled by trial and error learning to ingrain meaningful habits into their physical structures. But the physical effects of conscious feelings can be learned by trial and error only if conscious feelings have felt effects. This point is discussed in more detail in Stapp (2007).

3.3. Space-time structure

We now look in more detail at questions that orthodox quantum mechanics does *not* answer: What determines *when* a process 1 action will occur, and *what* the form of the associated physical action will be? That is, we turn to the problem of understanding the possible workings of process zero. These questions pertain to the representation of process 1 in space and time.

Von Neumann's analysis was based on non-relativistic quantum mechanics. According to orthodox non-relativistic quantum mechanics, each collapse event occurs at an instant of time, and it changes the state of the extended-over-all-space system just before time t to the state of the system at the instant of time *t*. The horizontal lines in the figure shown below represent the instants at which the state of the extended-in-space systems suddenly changes.

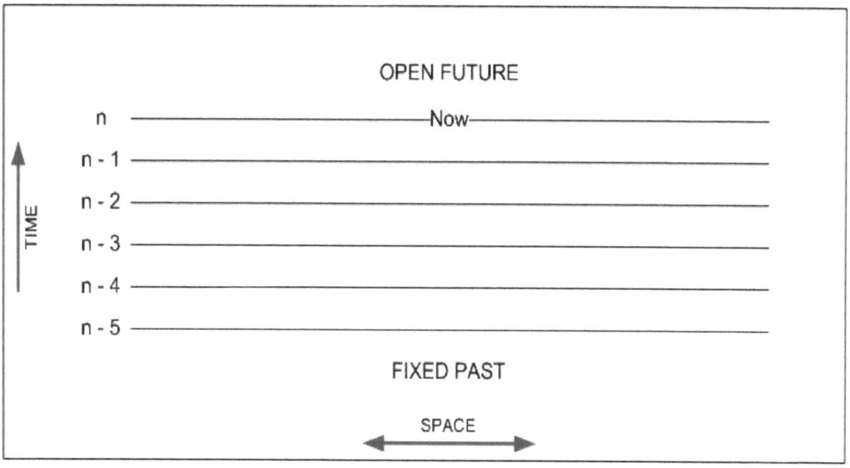

Fig. 5: A space-time diagram showing as horizontal lines the instants at which the evolving quantum state suddenly changes to a new (reduced) form. During the intervals between these times the state of the system evolves according to von Neumann's process 2, the Schroedinger equation.
(Source: Stapp, H. (2011). *Mindful Universe*. Berlin, Heidelberg, New York: Springer, p. 93)

Von Neumann's 1932 non-relativistic formulation was converted to a relativistic form during the middle of the century independently by S. Tomonaga (1946) and by J. Schwinger (1951). In this relativistic formulation the state of the system was associated not with an instant of time *t*, but rather with a space-like surface σ. A space-like surface σ is a continuous three-dimensional surface in space-time such that every point on the surface is space-like separated from every other point on the surface. A succession of collapse events can be assumed to occur on a succession of space-like surfaces σ such that each coincides with its predecessor except on a small patch, over which a surface σ is displaced slightly into the future relative to its predecessor, as indicated in figure 6.

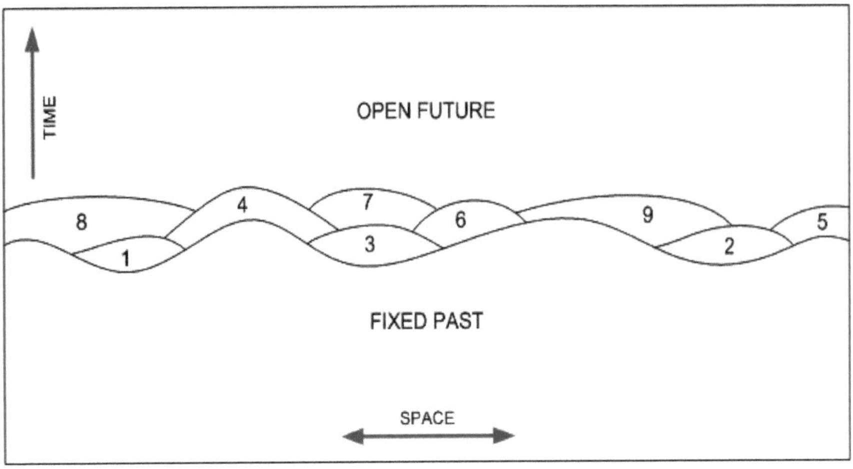

Fig. 6: The collapse events occur over a sequence of space-like surfaces σ each of which is locally shifted slightly forward in time relative to its predecessor. In the intervals between these surfaces the state ρ evolves in accordance with the relativistic quantum field theory (RQFT) generalization of the Schroedinger equation. (Source: Stapp, H. (2011). *Mindful Universe*. Berlin, Heidelberg, New York: Springer, p. 92)

The conceptual structure of the theory remains unchanged: the physically defined quantum state represents not "actuality" itself, but rather a set of objective tendencies pertaining to the occurrence of the next psychophysical event. It is these *events*, which are psychophysical entities, that are regarded as the objectively existing actualities.

3.4. Process time

The time represented by the vertical axis in figures 5 and 6 can be called "physical time": it is the time that appears very explicitly in our presently existing physical theories. But Relativistic Quantum Field Theory (RQFT) is naturally connected also another concept of time. This time has a sequence of intervals with the first interval associated with the stage of development of nature associated with the region 1 of figure 6, the next interval associated with the stage of development associated with region 2, and so on. I call this time by the name "process time" (Stapp 1985). This time permits the space-like surface σ upon which the *present* quantum state is defined to advance locally, as indicated in figure 6, rather than globally, as in figure 5. In particular, RQFT allows the quantum jumps associated with the various spacelike separated regions in figure 6 to occur in a definite sequence, marked by a sequence of instants in this *process time*. The orthodox rules of RQFT ensure that all *predictions of the theory* will be independent of the order in which *space-like separated events* occur in process time. It is this aspect of the theory that allows Relativistic Quantum Field Theory to be called "relativistic", even though the underlying process indicated in figure 6 has a preferred sequence of surfaces upon which the various quantum events occur. This occurrence of the actual reductions on the sequence of spacelike surfaces σ indicated in figure 6 accounts in a completely understandable way for the non-local aspects of quantum mechanics associated with the EPR (Einstein-Podolsky-Rosen) "paradox", and Bell's theorem.

To understand geometrically these non-local aspects of quantum mechanics one can represent process time by the coordinate along a fifth axis drawn perpendicularly to the four-dimensional plane indicated in figure 6. Then each instant in process time that is associated with a particular actual event will define a full four-dimensional space-time subspace in which lies the surface σ upon which the quantum state jumps to its new value. If one extends via process 2 the quantum state on this surface σ into the part of the full four-dimensional space-time that lies *earlier* than σ then, within this five dimensional space, one has a perfectly rational meaning for the assertion that each quantum collapse *changes the past*.

Within each of these four-dimensional subspaces all causal connections act into the forward light cones. The non-local aspects of quantum mechanics arise in this RQFT picture not from propagation of effects directly from one space-time region to a second one spacelike separated from the first. They arise rather from the condition imposed in one space-time region upon the entire physical space-time structure inherited from its process-time predecessor: the actualization of, for example, the potentiality for a particle to be in one particular region R, and in one particular spin state, eliminates from the prior space-time quantum state all those components that do not lead to this actualized location and spin state. This change associated with

R will *change the prior state* in a way that can change the potentialities for possible events located in regions space-like separated from the region R, thus producing an effective non-local effect.

The intervals in process time that lie between the instances that are associated with the actual events can be used to mark the steps in the process that determine what the next process 1 action will be.

For the discussion of such logical issues it seems, therefore, helpful to introduce this concept of a *well defined order of coming into being in a process time* that is *logically prior* to the less incisive *partial* orderings of events that occur in the classical Einstein theory of relativity, in which the order in which spacelike separated events occur is considered to be not only empirically undecidable, but – in keeping with certain positivistic precepts – also fundamentally undefined.

4. Connection to Whitehead

According to quantum mechanics, brain dynamics *at the microscopic scale* is subject to the uncertainty principle. This indefiniteness at the microscopic level tends to be magnified by the energy-releasing neurodynamics of the brain as one moves up to the *macroscopic* scale. Yet any gross indeterminacy at the macroscopic level conflicts with the empirical definiteness of our perceptions. A remedy is needed that will bring theory into alignment with experience.

The orthodox (Heisenberg/von Neumann) remedy is a theory built on the idea that actuality is built of psycho-physical events, with the evolving quantum physical state interpreted as both a repository of information and as a "*potential*" for the occurrence of the next actual event. Each process-1 psycho-physical event imbeds a correlate of a feature occurring in a stream of consciousness into the structure of the quantum mechanically described world, whereas each process-3 psychophysical event implants a correlate of a new piece of physical information into a stream of consciousness.

The idea of a reality built around localized psycho-physical actual events, and of "potentialities" for them to occur, is the core of a conception of reality advanced also by Alfred North Whitehead. He was stimulated by quantum mechanics, but constructed his theory of a psychophysical-event-based reality within the framework of the ideas of certain major figures of western philosophy; principally Plato, Aristotle, Leibniz, and William James. The central focus of Whitehead's work is precisely on the processes that I am calling "process zero", namely the process that formulates specific queries, and on the process that then chooses the answers to these queries, whose formulations and answers control the "creative advance into novelty".

The conception of the process of the unfolding of reality provided by RQFT meshes neatly with Whitehead's conception of the creative advance into novelty. One identifies each process 1 psycho-physical event together with its *logically subsequent* follow-up process 3 psychophysical event with a Whiteheadian actual occasion. This identification ties Whitehead's speculative philosophy into contemporary physics: it injects a science-based mathematical structure into the speculative philosophizing of Whitehead. On the other hand, orthodox quantum theory was specifically designed to be useful *to human beings*, and to be testable by them. Hence at that level quantum theory is manifestly anthropocentric. But within the Whiteheadian conception of process the human psychophysical events are considered to be mere special examples of the actual occasions that constitute the basic realities of the Whiteheadian creative advance into novelty. Whitehead's ideas provide therefore a philosophically grounded framework for enlarging the anthropocentric focus of orthodox quantum mechanics.

Although the quantum psychophysical reality is dualistic, in the sense that it includes mental and physical properties within an integrated causal structure, it is not Cartesian dualistic because every occurrence of any thought-like property is tied to a space-time region, and in specified ways to the physical properties located in that region. There is therefore a blurring of the Cartesian categories, with every individually existing property, both physical and mental, localized in a particular space-time region. Properties expressed in terms of *mathematically described understandings* become, then, not so radically different from properties expressed in terms of *psychologically described understandings*. This blurring of the mind-matter distinction constitutes a drift away from a completely *sharp* dualism, toward the "radical empiricism" of William James, or perhaps toward the "neutral monism" of Bertrand Russell (Russell 1945, 811).

Quantum mechanics was designed to account for the empirical regularities of our human streams of consciousness. Yet *something* was presumably going on before human beings arrived on the scene. So we are led to inquire about the nature of the quantum events that appear to lack the conscious dimension of the sort that we experience, but that appear to arise in purely mechanical ways. Indeed, William James, speaking about the flow of our own ideas says that "[n]o object can catch our attention except by the neural machinery. But the amount of attention that it receives after it has caught our attention is another matter. It often takes effort to keep mind upon it" (Stapp 2004, Sect. 12.7.4).

But how does quantum mechanics deal with the actions of the neural machinery, and more generally with apparent happenings at levels that are usually considered to lack any significant degree of mindfulness.

To better understand these "mechanical" events, it is useful to consider first the standard "interference" experiment shown in figure 7. Due to interference between the waves in the wave packets that follow the two alternative possible routes, every photon that enters at E is detected at detector B, none at detector D. (For later convenience, I have assumed that a glass bar has been inserted that shifts by 180 degrees the phase of the wave exiting the mirrors horizontally.) But if a 100% efficient, fully absorbing detector DID is inserted into either one of the two segments running between H_1 and H_2, then half of the photons that enter at E will be detected at DID, one quarter at B, and the other quarter at D. This distribution would be completely understandable if the wave packet of the individual photon goes wholly either one way or the other at H_1. Yet if DID were not inserted, then the wave packet for each individual photon entering at E would not be able to go wholly one way or another at H_1, for this would preclude the interference effect, and allow the photon to go often to D, which it never does in the case that DID is not inserted. But DID could be either inserted or not inserted according to a choice made randomly *only after* the wave packet has passed H_1. Yet how is the wave-packet for the individual photon able to decide whether or not to split at H_1 *before* the choice is made that determines whether or not it must split? And if the wave packet of the individual photon *always* splits at H_1 then how does inserting DID sometimes cause that photon to be deviated from B to D without the 100% efficient completely absorbing DID being affected in any way?

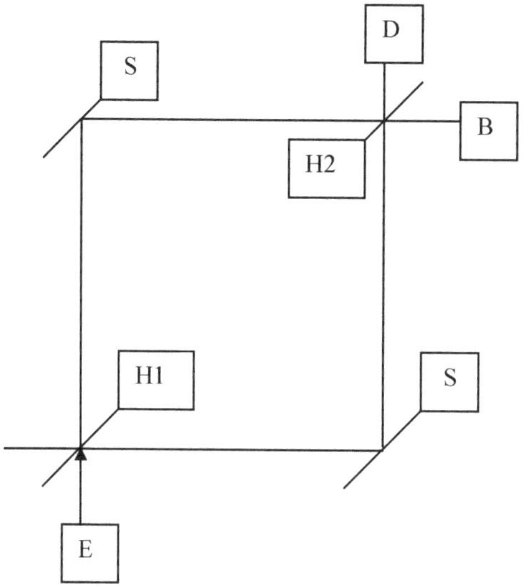

Fig. 7: A photon wave packet enters at E, and encounters a first half-silvered mirror H_1. Each part of the divided beam then encounters a fully silvered mirror S which diverts it to a second half-silvered mirror H_2. Two 100% efficient detectors, B and D, are placed in the two exit beams.

An even more puzzling/illuminating variation was devised by Kwait, Weinfurter, Herzog, and Zeilinger (1995). The experiment that they did is equivalent – as they note – to one that uses, instead of just two half-silvered mirrors, rather a sequence of N partially silvered mirrors each of which splits the beam *amplitude* into a reflected part that picks up a factor Cosine(π/2N) and a transmitted part that picks up a factor (i Sine(π/2N)). Both beams are then directed by mirrors through the next partially silvered mirror, giving a generalization of figure 7, with N partially silvered mirrors P_1, P_2, … P_N instead of just the two half-silvered mirrors H_1 and H_2. In complete analogy to figure 7 the beam that vertically exits any P_J, with J = 1, 2, … N-1 is reflected so as to enter $P_{(J+1)}$ horizontally, and vice versa. The final vertical and horizontal beams go to detectors D and B, respectively. One again finds that *if no internal detector is put into the system* then one of the two external detectors, namely B, will certainly fire, and other external detector, namely D, will certainly not fire. But suppose, for large N, that a 100% efficient, and completely destructive, detector (a DID) is placed in each *upper* path between two P_J s. Then, for *almost every photon* that enters the apparatus, the detector D will fire, instead of B, *and none of the inserted DIDs will fire*. Thus *almost every photon* that enters the system will be diverted, *completely intact* (i.e. with all of its energy) from B to D by inserting this set internal detectors DID, *all* of which will left *completely unaffected.*

These effects are all accurately predicted by quantum theory, as logical consequences of the effective backward in time effect of the actions of the detectors that was described in the previous section. The quantum state represents potentialities, and each actualization can be considered to change potentialities in the entire space-time region leading up to the surface σ upon which the collapse occurred.

According to the *classical* wave mechanical theory, there would be, for any finite N, and for *every* entering pulse, *no* possibility that even one DID would be left completely unaffected, whereas according to quantum theory, for a large N, *every* DID will, for almost every entering photon pulse, be left *completely* unaffected. According to the classical wave mechanical theory, with all the DIDs in place, *every* input pulse will emerge with less energy than it brought in, whereas according to quantum theory every pulse that exits the system will retain all of the energy that it brought in. The two theories agree, on the average, but at the level of the individual quantum-sized pulses they are very different, because each entering quantum photon is detected with all of its energy at one and only one of the N detectors.

The fact that for large finite N, with the DIDs in place, there is a large probability that the passage of the photon will leave each DID *completely unaffected* is a consequence of the quantum Zeno effect: the finely spaced "measurements" tend to keep DIDs *exactly* in their original unexcited state. The orthodox way to understand the quantum dynamics – in a way that goes beyond merely understanding how to apply the practical rules – is to understand that the actual things are *events located in regions*, not moving objects. Placing detectors in region R *modifies that region.* It creates in R a potentiality for a certain kind of events either to occur there or not to occur there. For large N, there is a large likelihood that *no actualization event* occurs in the region R occupied by the DIDs, but that, nevertheless, the photon energy is, by virtue of the change made in the region R, forced to go to D instead of B. This conception of reality in terms of *properties of regions to actualize or not actualize particular potentialities*, in strict accordance with the statistical rules of quantum theory, is the core idea of the Quantum-Whiteheadian approach. If the potentiality for a photon to be found in one region is eliminated by an event occurring there, then the potentiality for that photon to be found elsewhere is immediately increased: there is a *nonlocal quantum jump.*

The example discussed above shows how, with a fairly high level of reliability, a fairly large amount of energy can, by changing the properties of a region R, be diverted from B to D, *without any effect at all* upon either the amount of energy delivered to D at a detection event that occurs there, or upon the character of the region R whose altered properties produced the diversion from B to D. In the classical analog each pulse arriving at D will have less energy than what it originally had, and its passage will have left the properties of R slightly altered.

Classical mechanics gives a deterministic description of how the energy gets distributed on the average, whereas quantum mechanics gives a statistical description of how the energy gets distributed in the individual instances. Reality, at least as we experience it is made from individual instances, so something is missing from a theory that treats the averages as the basic reality, namely the process that determines what actually happens.

Insofar as the fundamentally quantum mechanical brain can evolve in a way that can usefully exploit these potentially available quantum mechanical features one might reasonably expect it to do so. Such a brain could be superior to a classical brain in the trial and error development of a repertory of useful templates for action: the fine tuning of transmitted energies and of properties of the medium can be important to the optimal functioning of a complex dynamical quantum system. (Consider the transmission of information via a photon emitted by an atom at one site and absorbed by the same kind of atom at another site) Also, "[t]he quantum dynamics allows "optimal" self generating resonant states to emerge from the amorphous quantum soup with a certain maximal efficiency because all of the possible overlapping configurations of classical possibilities are simultaneously present, and their consequences are simultaneously explored by the quantum dynamics" (Stapp 1994, 288). Here "optimal" means most likely to lead to the intended feedback. A reduction of the dynamics to a more classical dynamical level tends to disrupt the stability of the quantum-stabilized state (ibid).

According to Whitehead, each actual occasion must have some mental aspect. But the physical examples described above indicate that the laws of nature are such that certain regions can be treated as the loci of actual quantum detection events, even though the physical properties located in those regions would not warrant assigning any appreciable amount of conscious awareness to these region themselves.

The condition that, at some level, a particular detection event definitely either happens or does not happen (in accordance with the quantum probability rule) requires *at that level* some sort of discrete selection of what the physical event is that will either happen or not happen. There is no empirical evidence for the entry of any element of discreteness before the entry of conscious awareness. It is therefore logically possible to maintain a purely mechanical determinism at all non-conscious levels of the universe by assuming, in accordance with both orthodox quantum mechanics and Whiteheadian philosophy, that any entry of a discrete choice occurs only in association with an aspect of nature having the character of conscious awareness (increase in knowledge), and to allow wave mechanical deterministic continuity to prevail in all mindless activity. That is, there is no *logical need* to introduce the notion of a *discrete choice* of one particular mechanically generated branch, from among a continuum of possibilities, before introducing the notion of a consciously aware mind, or the associated notion of an evolving body of knowledge.

On the other hand, as emphasized by the examples described above, quantum mechanics also *allows* regions presumably not themselves supportive of conscious awareness to be treated *as if* they were loci of actual events/occasions, insofar as some future experience will effectively depend upon the choice specified by that contemplated actual event. This moveability of the level at which the actualization events can be considered to occur was often emphasized by the founders of quantum mechanics, and was effectively demonstrated by von Neumann (1932/1955) in his theory of the process of "measurement".

Quantum theory does not, however, rule out the possibility that 1), there are some – as yet unknown – essentially mechanistic laws that cause, without any reference to consciousness, certain process 1 events to occur automatically; that 2), only certain sufficiently complex feedback events, associated for example with biological systems, will have a conscious/non-mechanical component; that 3), this conscious/non-mechanical component has causal efficacy by virtue of its capacity to rapidly repose the process 1 query associated with the feedback; and that 4), this causal role of the conscious/non-mechanical aspect is not reducible to the mechanical realm because the decision to repose is related to "feelings" whose power to repose the query may not be causally determined by the physical state alone This arrangement would be in line with the quote of William James cited above, which claimed that certain initial events are generated by the neural machinery process alone, and would allow the quantum mechanical actualization process to proceed in the absence of consciousness. But it would impossible to demonstrate empirically that such events have no mental aspect whatsoever. Hence the distinction between this theory and Whitehead's would be a matter of philosophical preference.

5. Classicality and brain process.

The question arises: Why should a quantum mechanically described brain produce classically describable conceptions of the physical world?

The quantum mechanical brain can be presumed to be represented by quantum electrodynamics, which deals with relationships between the motions of charged particles and changes of the electromagnetic field. This field has a class of states called "coherent states" that have properties very similar to the properties of corresponding states of classical electromagnetic fields (Klauder 1968, 1985). They can be defined by averages over regions that are large compared to atoms, but are small compared to the whole brain. These states tend to be dynamically stable, and to obey the dynamical equations of classical electrodynamics. This stability makes the reductions to these states the ideal candidates for the physical aspects of those events whose mental aspects populate our streams of consciousness (Stapp 1987, 1994).

This likely connection between our conscious thoughts and these quasi-classical states of the electromagnetic fields in our brains links naturally to the notion that the neural correlates of our conscious thoughts are normally associated with the 1 to 50 Hertz oscillations found in our brains.

The logical continuation of the this discourse on the effect of mind upon brain involves this proposed connection between the 1 to 50 Hertz brain waves and these quasi-classical states of the electromagnetic fields in our brains. This connection is discussed in three chapters in recent books (Stapp 2009 a, b, c). They focus particularly on the technical details of the bodily implementation of conscious intent through application of the quantum Zeno effect.

References

Bohr, N. (1958). *Atomic Physics and Human Knowledge*. New York: Wiley.

Heisenberg, W. (1958a). "The Representation of Nature in Contemporary Physics". In: *Daedalus*, 87 (summer), pp. 95-108.

—— (1958b). *Physics and Philosophy*. New York: Harper.

Kwait, P.; Weinfurter, H.; Herzog, T.; Zeilinger, A. (1995). "Interaction-Free Measurement". In: *Phys. Rev. Letters*, 74, pp. 4763-66.

Neumann, J. von (1955). *Mathematical Foundations of Quantum Mechanics*. Princeton N. J.: Princeton University Press (translated from the Springer 1932 German original by Robert T. Beyers).

Russell, B. (1945). *History of Western Philosophy*. New York: Simon and Schuster.

Schwinger, J. (1951). "Theory of Quantized Fields I". In: *Physical Review*, 82, pp. 914-927.

Stapp, H. (2009a). "Physicalism Versus Quantum Mechanics". In: *Mind, Matter, and Quantum Mechanics* (third edition). Berlin, Heidelberg, New York: Springer, (Chapter 13) pp. 245-260.

—— (2009b). "A Model of the Quantum-Classical and Mind-Brain Connections, and the Role of the Quantum Zeno Effect in the Physical Implementation of Conscious Intent". In: *Mind, Matter, and Quantum Mechanics* (third edition). Berlin, Heidelberg, New York: Springer, (Chapter 14) pp. 261-276.

—— (2009c). "Philosophy of Mind and the Problem of Free Will in the Light of Quantum Mechanics". In: von Mueller, A. (ed.). *On Thinking*. Berlin, Heidelberg, New York: Springer (arxiv.org/abs/0805.1633.0116).

—— (2007). *Mindful Universe: Quantum Mechanics and the Participating Observer*. Berlin, Heidelberg, New York: Springer.

—— (2004). *Mind, Matter, and Quantum Mechanics* (second edition). Berlin, Heidelberg, New York: Springer.

—— (1994). "Quantum Mechanical Coherence, Resonance, and Mind". In: *Proceedings of Symposia in Applied Mathematics, Volume 52, Proceedings of the Norbert Wiener Centenary Congress.* American Mathematical Society.

—— (1985). "Einstein Time and Process Time". In: Griffin, D. R. (Ed.). *Physics and the Ultimate Significance of Time*. Albany: Suny Press, pp. 264-270.

Tomonaga, S. (1946). "On a Relativistically Invariant Formulation of the Quantum Theory of Wave Fields". In: *Progress of Theoretical* Physics, 1, pp. 27-42.

A Fourth Variable in Evolution

JOHN B. COBB, JR.

I cannot contribute to the discussion of evolution as a scientist, and I cannot claim to have read the literature widely or thoroughly. Accordingly, I am only reporting as a Whiteheadian on what I have learned through a conference on evolution and religion and by working on the papers to develop them into a book.[1]

Most of the conferences on science and religion take these two areas of thought as more or less given and then consider how they relate. Our Whiteheadian approach is quite different. We assume that revisions are needed on both sides. I thought that it would be appropriate in the context of this section to report on my conclusions as to the sort of revision that is needed on the side of evolutionary theory.

1. The activity of organisms as a main factor in evolution

The evolutionary theory that requires revision, I call neo-Darwinism. It is so widely accepted, that most biologists feel no need to give it a special label. At our conference the strongest representative of this mainstream theory was Francisco Ayala. He agreed to write additional material for our book so that the thinking of the majority of working biologists would be well represented. The book as a whole, however, is an argument against the adequacy of his formulation of evolutionary theory.

---

[1] The Center for Process Studies received a small grant from the Metanexus Foundation, which emphasizes discussion of the relation of science and religion. We have held three conferences. The first conference, on evolution and religion, took place in October 2004. I had primary responsibility for organizing it and also for developing the book *Back to Darwin* from it.

Ayala is not a neo-Darwinian in the full sense. He believes that there are ways of knowing other than science. Accordingly, he does not draw from his evolutionary theory the full range of neo-Darwinian conclusions. He represents a widespread way in which peace is established between science and religion, that is, an epistemological dualism. However, I am not reporting here on the ways of dealing with the relation of science and religion – only on the science. I begin by quoting two passages from Ayala, so that you will know exactly what position I want to revise.

> "I argue in this paper that science encompasses all of reality and that we owe this universality to Charles Darwin, who completed the Copernican Revolution by extending to the realm of life the Copernican postulate of the natural world as matter in motion governed by natural laws. The Copernican Revolution had left out the diversity and configuration of organisms, because organisms and their parts manifest to be designed. Natural selection acting on spontaneously arising mutations can account for the diversity of organisms and their design. Evolution is a creative process owing to a fruitful conjunction of contingent and deterministic processes" (2008a, 50).
>
> "Natural selection does not strive to produce predetermined kinds of organisms, but only organisms that are adapted to their present environments. Which characteristics will be selected depends on which variations happen to be present at a given time in a given place. This in turn depends on the random process of mutation (broadly understood), as well as on the previous history of the organisms (i.e., on the genetic make-up they have as a consequence of their previous evolution). Natural selection is an 'opportunistic' process. The variables determining in what direction it will go are the environment, the preexisting constitution of the organisms, and the randomly arising mutations" (ibid. 72f.).

It is my impression that these quotations express views widely held by contemporary evolutionary biologists. They are quite standard formulations. They lead to the conclusion that everything is to be explained by chance and necessity. Actually, the "chance" is usually not thought to be, finally, pure chance, since it is assumed that physical laws apply. These may be regarded as statistical laws, allowing for some small element of indeterminacy, but this is not what is meant when mutations are said to be random or matters of chance. The point is instead that mutation is not in any way purposeful or influenced by purpose, divine, human, or natural.

As a Whiteheadian, I find this explanation of evolution implausible, not because of what it includes but because of what it omits. It includes the organisms, understood as the multicellular plants and animals. It includes the environment of these organisms. And it includes the genes.

It omits the environment of the genes in the DNA molecule, the environment of that molecule in the cell, and the environment of the cell in the body. Instead it speaks of the randomly mutating gene as unaffected by its environment. Except for the selection of genetic mutations for survival, the standard theory omits the influence of the environment on the phenotype, that is, the multicellular organism. It omits the effects of changes in the phenotype in determining the adaptiveness of mutations. And finally, it omits the activity of the multicellular organisms and therefore the influence of this activity in determining both the adaptiveness of genetic mutations and the environment that selects.

If we consider the environment of genes to be composed of tiny organisms, then we can regard most of what is omitted as consisting of the activities of organisms at many levels. To simplify, I will emphasize here the activity of multicellular organisms, but in fact eucaryotic single-celled organisms, bacteria, viruses, and proteins are also important factors in evolution. I intend for my fourth variable to include all these even though I do not develop the argument here.

Mainstream biologists know that there is evidence for all of the relationships they omit from their formulations of the mechanics of evolution. Ayala refers to much of this. The problem is not ignorance. My complaint is that, despite his rich knowledge and that of other biologists, as a group they see no need to include any of this in their standard explanations of evolution. In this short paper I will review some of the evidence and consider the reasons for its neglect.

The phenomenon most widely recognized by evolutionary theorists that points directly to the role of the activity of multicellular organisms, in this case animals, is called the Baldwin effect or genetic assimilation. Baldwin showed that there is considerable evidence that animals find new adaptive modes of behavior and that these lead to the selection by the environment of different phenotypes, and through them, eventually, of genes. Thus animal activity not determined by genes influences nature's selection among the randomly mutating genes. It functions as a variable alongside those identified by Ayala. Although this is widely acknowledged, it is generally regarded as a minor factor, and it is ignored in more summaries of the causes of evolution.

I was curious to learn how Ayala would respond to this, and I appreciated his writing a section on the Baldwin effect for the book. He wrote that this effect is fully recognized by neo-Darwinians and has been integrated into their theory. This has been possible, I gather, because his account of the Baldwin effect does not include any reference to the activity of organisms. I quote again:

> "Simply stated, the hypothesis asserts that the environment affects adaptively the phenotype of an organism, that is, its configuration, so that the organism can survive and reproduce under conditions that are unusual or even extreme, and that such adaptive modifications may later become genetically fixed by natural selection" (Ayala 2008b, 193).

I was struck by two things. First, Ayala seems to accept the idea that the priority of the effect of the environment's influence on the organism is widespread rather than occasional. This is quite different from the usual picture, in which selection of adaptive mutations of genes is not mediated by prior changes in the phenotype. I quote the paragraph with which Ayala concludes his discussion of the Baldwin effect.

> "The Baldwin effect has been ascertained in many other instances, including cast-determination in social insects (ants, termites and honeybees) and the affinity of hemoglobin for oxygen. Indeed, the Baldwin effect has been generally involved in the origin of evolutionary novelties.[3] Evolutionary novelties are reorganizations of preexisting phenotypes, which first arise in response to environmental challenges (given that all genotypes have enormous plasticity, that is, a wide norm of reaction), but eventually become genetically determined if the particular environmental challenges persist and the adaptation importantly contributes to survival and reproductive success. Kirschner and Gerhart explain in considerable detail that the genetic changes that account for evolutionary novelty involve gene control circuits (numerous but very short DNA sequences), rather than changes in the enzymes encoded by genes" (ibid. 195).

If I understand this correctly, then, contrary to the impression given by most evolutionists when speaking to the public, changes of phenotypes in response to environmental conditions are a *major* factor in evolution. Subsequent selection among random genetic mutations functions to stabilize changes that have arisen in other ways. The Baldwin effect, instead of being a peripheral oddity, is of enormous importance for evolution. If so, evolutionary theory and its presentation to the public should be reformulated to make that clear. Instead of the environment selecting those randomly mutated genes that happen to produce phenotypes with survival advantage, the selection is of genes that support changes in the phenotype that have previously been selected for their superior adaptiveness.

An even more radical priority of phenotypes over genes seems probable. Among contributors to our conference and to the book developed from it is Dorion Sagan. He emphasizes the physical basis for evolutionary phenomena and with respect to the origin of genes he writes as follows:

> "Genes themselves likely appeared secondarily, after metabolically sustaining energy systems. It was their ability to further stabilize already relatively continuous centers of energy degradation, favored without natural selection by the second law that allowed genes the chance to arise in the first place. Consider an effective replicator hobbling its reproductive success by growing a great, protein-studded body. Why would such a replicator shoot itself in the foot? Non-metabolizing replicators would easily replace it. This thought experiment proves that thermodynamic complexity, in the form of relatively stable but not yet fully replicative metabolizing bodies, must have appeared before naked DNA, or the (popularly postulated) RNA world" (Sagan 2008, 148f.).

In an article entitled "A Unified View of the Gene, or How to Overcome Reductionism", Peter Beurton provides an extended account of the emergence of genes in biological terms. He summarizes thus: "Not only are the selective values of genes emergent properties, but the genes *themselves* are emergent particles resulting from the interactive processes of populations" (2000). In short genes came into being through the Baldwin effect.

Second, I was also struck by the fact that much of this can be acknowledged without giving up what neo-Darwinians consider central to their theory. This is that the only agents in evolution are the genes and the environment. What can be assimilated by neo-Darwinists, despite its invisibility in their formulations, is that the effect of the environment on the phenotype precedes genetic change. This leaves the genes and the environment as the active agents, with the preexisting condition of the organism as the object of their action. However, the assimilation of the Baldwin effect into neo-Darwinism is only partial. The evidence for the operation of the Baldwin effect where the activity of organisms is a causal factor is ignored.

There are then two issues with neo-Darwinism. First, if the evidence now indicates that the Baldwin effect, broadly understood, identifies the primary pattern of evolutionary change, the standard explanation of evolution should make this visible. It does not suffice to assure the public that this effect can be interpreted in neo-Darwinian terms, meaning that it can be explained without adding any third agent of activity to the genes and the wider environment.

Second, the evidence that the *activity* of multicellular organisms plays a role in at least some instances of the Baldwin effect should also be directly considered rather than ignored. It is hard to see how one can deny the role of animal activity altogether. The codling moth once laid its eggs only on apples. But some codlin moths began laying their eggs on walnuts. These have developed into a distinct species. However one explains this change in host selection, I do not see how one can avoid the fact that the activity of moths was involved. Ignoring such facts is not a good scientific way of maintaining a theory that cannot account for them. Genetically unprogrammed activity of organisms *is* a distinct variable explanatory of evolutionary developments. Yet the prejudice against including the activity of organism as a variable in directing the course of evolution is so great that even sexual selection is rarely mentioned.

This role of the activity of organisms in indirectly determining the selection among genes is complemented by its role in modifying the environment. Richard Lewontin has recently emphasized the importance of this activity and complained about its neglect by evolutionists (2000). That neglect encourages, and is encouraged by, its omission from evolutionary theory. We saw that the quote from Ayala above, as well as his account of the Baldwin effect, depicts organisms as passive in relation to their environment as well as to the genes. But it cannot actually be doubted that organisms affect their environments. Further, the changes they effect in their environment certainly also affect what is adaptive both for themselves and for other organisms.

This is so obvious that it seems unnecessary to illustrate it. If a new predator enters an ecosystem, it certainly changes the environment for its prey. Habits that may previously have been adaptive may no longer be so.

Indeed, the effects of the activity of organisms go far beyond that. Billions of years ago, organisms that generated oxygen made the atmosphere poisonous for many creatures and radically redirected evolutionary development. There is now considerable evidence that the biosphere as a whole regulates the climate of the planet. These are surely not minor matters easily relegated to the periphery of theory.

Symbiosis describes an activity of organisms that dramatically changes their environments. Symbiotic relations lead to genetic changes. Ayala does not question that there is lateral transfer of genes as a result of such relations. The emergence of the eucaryotic cell, perhaps the single most important event in biological evolution, is now recognized as an instance of symbiogenesis rather than of mutation of genes.

I hope that these brief comments are sufficient to convince anyone who pays attention to what is going on in the world, that the activity of organisms plays a distinct and important role in evolution. Neo-Darwinists cannot realistically deny this, but if they cannot, then continuing to omit this activity from their accounts of evolution is surely unjustified. Hence, my thesis is quite simply that alongside the three variables listed by Ayala: the genes, the environment, and the extant state of the organism, we need a fourth: the *activity of organisms*.

2. Organisms as experiencing subjects

If there is as much evidence as I believe for the role of this activity, we must ask why it is systematically neglected by evolutionists. I suggest that we can find the answer in a now famous quote from the biologist who, himself, protested the neglect of this activity: Richard Lewontin.

> "Our willingness to accept scientific claims that are against common sense is the key to an understanding of the real struggle between science and the supernatural. We take the side of science […] because we have a prior commitment, a commitment to materialism. It is not that the methods and institutions of science somehow

> compel us to accept a material explanation of the phenomenal world, but, on the contrary, we are forced by our a priori adherence to material causes to create an apparatus of investigation and a set of concepts that produce material explanations, no matter how counterintuitive, no matter how mystifying to the uninitiated. Moreover, that materialism is absolute, for we cannot allow a Divine Foot in the door" (1997).

For Lewontin, as for many biologists, this means that we cannot allow subjectivity of any kind to play an explanatory role. From his point of view, since nature is purely material, subjective experience is "supernatural". Lewontin thinks that he can affirm the importance of the activity of organisms and still exclude any role for animal experience. However, most biologists recognize that this proves difficult. Organisms seem to act so as to obtain food, to be safe, and to have pleasure. This seems to be because of their subjective desires. Given their sense organs it is hard to doubt that they have visual and auditory experiences and that their actions are influenced by these. It is hard to doubt that their actions are purposive. Once we allow that animal activity significantly influences the course of evolution, it is difficult to exclude a role for animal purposes. Rather than allow an important role for animal activity and then argue that it is exhaustively explained by randomness and law, chance and necessity, most biologists prefer to ignore the activity even at the expense of leaving out of consideration a great deal of empirical evidence.

Lewontin is clear that he argues as he does because of his metaphysical commitment. Would that neo-Darwinists in general were that honest! We may respect such honesty and such commitment, but it leads to a shift in the debate. If science is understood to be a dispassionate search for truth, then science does not support neo-Darwinism. If science is limited to acknowledging only what fits with its *a priori* metaphysics, neo-Darwinism may be good "science," but there is then no reason to privilege "science" over other ways of interpreting the evidence.

In 1938 Whitehead addressed this question perceptively:

> "Science can find no individual enjoyment in nature: Science can find no aim in nature: science can find no creativity in nature; it finds rules of succession. These negations are true of Natural Science. They are inherent in its methodology. The reason for this blindness of Physical Science lies in the fact that such Science only

> deals with half the evidence provided by human experience. It divides the seamless coat – or, to change the metaphor into a happier form, it examines the coat, which is superficial, and neglects the body which is fundamental" (211).

For Whitehead subjective experience and what is perceived objectively belong together in a single world. The actual course of evolution is a product of both working together. There is extensive evidence for this. The problem is a view of science that excludes the subjective from its purview. The problem is greatly intensified when scientists, have excluded the subjective, still claim that their explanations, based only on what can be perceived objectively, that is, by an outsider, are complete.

Obviously, it is my view that the adoption of a Whiteheadian metaphysics enables one to accommodate the evidence more inclusively. It is also my view that more scientists should be willing to allow the evidence to trump their metaphysical commitments. If in one way or another, the activity of organisms is recognized as an important factor in evolution, then it becomes possible to view human beings as part of the evolutionary process without supposing that we are nothing but matter in motion.

Honest scientists, in my view, have two choices. They can continue to restrict themselves to a methodology that limits science to the study of what is purely objective and then recognize that there are other factors at work that they cannot include. This means that they will acknowledge that scientific explanations are inherently incomplete. Or they can expand the study of science to include subjective aspects of the nature they study. They will then be able to include the purposive character of living things in their explanations.

I will conclude by mentioning one example of what has happened when science is expanded in this way. I should acknowledge that this example has no direct relation to evolutionary theory.

Charles Hartshorne, a leading Whiteheadian philosopher, was also an ornithologist. His special interest was birdsong. The usual explanation of birdsong is that birds sing only to attract mates and defend territory. Hartshorne believed that, in addition to these purposes, birds sing because they enjoy singing.

He wanted to test this hypothesis. He reasoned that if he was correct, birds with very simple songs would wait longer between songs. Quick repetition would be boring. On the other hand, birds with more elaborate repertoires would pause less. He gathered extensive evidence, and it strongly supported his hypothesis. His book, *Born to Sing*, has had a good reception among ornithologists. Allowing for investigations of this kind does not destroy science, as some scientists, such as Lewontin, seem to fear.

References

Ayala, F. (2008a). "From Paley to Darwin: Design to Natural Selection". In: Cobb Jr., J. B. (ed.). *Back to Darwin: A Richer Account of Evolution*. Grands Rapids, MI: Eerdmans Publishing Company, pp. 50-75.

—— (2008b). "The Baldwin Effect". In: Cobb Jr., J. B. (ed.). *Back to Darwin: A Richer Account of Evolution*. Grands Rapids, MI: Eerdmans Publishing Company, pp. 193-195.

Beurton, P. (2000). "A Unified View of the Gene, or How to Overcome Reductionism". In: Beurton, P.; Falk, R.; Rheinberger H. J. (eds.) (2000). *The Concept of the Gene in Development and Evolution*. Cambridge: Cambridge University Press, pp. 286-314.

Lewontin, R. (2000). *The Triple Helix: Gene, Organism, and Environment*. Cambridge Mass.: Harvard University Press.

—— (1997). "Billions and Billions of Demons". In: *New York Review of Books*, 9, pp. 28-32.

Sagan, D. (2008). "Evolution, Complexity, and Energy Flow". In: Cobb Jr., J. B. (ed.). *Back to Darwin: A Richer Account of Evolution*. Grands Rapids, MI: Eerdmans Publishing Company, pp. 145-156.

Whitehead, A. (1938). *Modes of Thought*. New York: Macmillian.

No Need for Dualism in Evolutionary Theory. A Comment on John B. Cobb's "A Fourth Variable in Evolution"

ANDREW PACKARD

I welcome this opportunity to comment on John B. Cobb's essay. Our editor has asked me to write for philosophers and scholars who are interested in philosophical problems of life rather than writing for philosophers of biology.

That will not be easy. Cobb is tackling aspects of the philosophy of biology and since I am a biologist, though of a generation now mostly retired, whatever I say will need to stand the test of other biologists looking over my shoulder. More importantly it will need to be true to my own experience of living entities.

1. Reflections on Cobb's criticism of a fragmented science

1.1. Cobb's fourth variable is a Trojan horse

When Cobb writes that he finds Francisco Ayala's "explanation of evolution implausible, not because of what it includes[1] but because of what it omits" he is implying that *technical formulations of evolution suffer from the same limitations as technocratic solutions to problems in the real world.* His Fourth Variable[2] exposes weaknesses in the Neo-Darwinian edifice and of all attempts to embrace within it a guiding role for the activities of organisms – if only because of the unpredictable nature of the phe-

[1] See Ayala's summary [in Cobb's chapter] this volume.

[2] Cobb's Fourth Variable is a set of other relationships contributing to the mechanics of evolution in the real world. It includes activities of the phenotype covered by the so-called "Baldwin Effect".

notype. Simple appeals to random mutation or other single factor certainly can not bear the weight the "Baldwin effect" would put upon them.

Spyridon Koutroufinis (this volume) describes some of the problems of modelling self-organising systems without at the same time pre-setting the many variables that in life can dramatically alter the behaviour of the system.[3]

In a review of the (confusingly named) "Baldwin Effect" and the issues it raises for Darwinian evolution, Patrick Bateson (2004) examines four proposals for "ways in which an animal's behaviour could affect subsequent evolution".

The debate on whether or not there can be laws in ecology and what part mathematical models play in ecological theory is also relevant here (see Cooper 2003).

1.2. Evolutionary theory and practice of most professional biologists essentially uncoupled

There is an enormous literature in evolutionary epistemology.[4] Some of it considers the issues raised by Cobb. This literature is better known to philosophers, and to students doing courses in philosophy of biology, than to most biologists. Evolutionary theory and the practice of "most biologists"

---

[3] "It is typical of all mathematical accounts of self-organized behaviour, with which I am familiar – whether in physics, chemistry, or biology – to *essentially depend on a high number of externally set parameters*" Koutroufinis (this volume, see sections 1.4 and 1.3).

[4] From this large literature, I mention the following: Gregory Cooper (1988) examines the philosophical status of Neo-Darwinism; Erkki Haukioja (1982, see below); Gabriel Dover (2000) "exposes the naively deterministic view of selfish genes" from within the establishment of modern molecular genetics; Robert G.B. Reid's innovative evolution (Reid 2007) substitutes self-amplifying "natural experiment" for natural selection – taking place under the "big top" of a 3-ringed circus where development, physiology and behaviour, environment, continually interplay. Many of the authors of this literature, and most of the public they are writing for, are not biologists. Biologists constitute only 5 of the 25 essayists contributing to a recent book (Grafen and Ridley, 2006) celebrating 30 years since the publication of Dawkins's *Selfish Gene*. Authors are exclusively from the USA, UK, Canada, Australia and New Zealand.

are on different planes steeply inclined to one another.[5] Practising biologists are rarely if ever required to submit their findings to any litmus test of Neo-Darwinism – or vice versa. This does not stop biologists referring to theory when writing up results – often in an uncritical and politically correct lip service to a highly reduced version. We tend to allude, in the Discussion section of our papers, to the potential (but not objectively demonstrated) "selective advantage" of a particular process, or trait, that we have examined. Less innocently, when experimental observations appear to be at variance with theory, authors can resort to all kinds of intellectual contortions in order to remain faithful to prevailing *cultural expectations* that have no connection with the science.[6]

1.3. Ratiocination and the technical expectations of society

A comment of Alfred North Whitehead in conversation with Lucien Price of the *Boston Globe* can perhaps help us see what is happening here. "Many of the people, including prominent ones, who are now regarded as scientists are little more than technicians" (Price 1954, 67).

[5] Fitness equations have a short-term *future* projection which can rarely if ever be tested against facts of evolution *past and present*. Most of the Neo-Darwinian literature is *self-referent*. Gregory Cooper (1988), quoting Alexander Rosenberg, points out that "the biological fields which typically study the ways in which selection pressures originate in the organism/environment complex are not in fact part of the theory of natural selection at all".

[6] The entrenched nature of a cultural assumption ("self-interest") in certain quarters, and of its role in evolutionary theory, is illustrated by a recent scientific article on cooperation among microorganisms (Wingreen and Levin 2006). Opening with the statement: "One of the organizing principles of life on earth is that cells cooperate", the Discussion section then seeks to explain the principle by reference to (effectively opposite) anthropomorphic ones unrelated to the findings: "Is cooperation best understood as the convergence of the immediate self-interest of multiple parties? Or can evolution lead to stable cases of short-term altruistic behaviour, providing long-term benefit for all?" For a brief list of authors (from Empedocles to Wheeler) who have treated cooperation as fundamental principle in nature, see W.C. Allee (1938, Chapter 2). In Packard (2006), I summon basic physiological facts in an attempt to remedy its curious neglect in modern biology. However, facts alone can not counter the force of indoctrination over reasoned argument.

Seventy years on, we have all become technicians. Specialisation distances us from each other and from the understanding of the whole. The age is increasingly dominated by sophisticated (and often expensive) techniques of one kind or another. Despite its welcome emphasis on process, even much-trumpeted *Systems Biology* is yoked to *biotechnology*: to exploiting the processes and accompanying desire to *control.* The persistent danger is that students and workers in the Life Sciences, in line with the technological expectations of society and its cultural distortions, fail to recognize that technocratic solutions can only solve technical problems. To be successful such solutions must be complete. Simulations of an overarching theory of evolution suffer the same limitation. They can never be satisfactorily complete.[7] Life is too elaborate.

To recover a proper view of the self-regulatory whole, which accords due place to its intrinsic controls, will require a major programme of re-education.

1.4. Troubles with the (Anglo-American) school of evolutionary biology

The (Anglo-American) school of Neo-Darwinism is largely concerned with the science of prediction – notably with modelling population dynamics and the prospective fate of genes, traits and behavioural stratagems – requiring not inconsiderable knowledge of one kind of mathematics.[8] Moreover, some of the apologists for current formulations infer that biology (a largely descriptive science) can be considered mature and professional

[7] Tragically, as we know from the rates at which marine ecosystems are being degraded (by shrimp and salmon farming practices, for instance (Molyneaux 2007)), the inverse is equally true. Technical solutions applied to the complex fruits of evolution can never completely reflect biological realities.

[8] Darwinism itself is now called a science. Student entry requirements for certain courses of evolutionary biology include the mathematics of Neo-Darwinism. The well is deep and can fill a lifetime. However, recent enormous strides in understanding the *activities of genes in the environment of the genome* have made some of the assumptions of the fundamental fitness equations look decidedly silly (see Dover 2000). Reid (2007) writes of the "mindset that regards evolution as no more than the differential reproduction of genes".

only if it has theories and laws that can be expressed mathematically: that it is better to have simplicity and clarity – even though the theory may be erroneous, flawed, or not apply in the case under consideration – than *not* to have a law or a theory. Sometimes even the best of these apologists trip themselves up on the skirts of their own convictions. In *Darwinism and its Discontents* (Ruse 2006), the historian and philosopher of Darwinism Michael Ruse traces the 20th century development of "a proper and professional evolutionary theory" and its relation to philosophy. The book has something of the grandeur of a Faustian struggle for the soul. A runaway – perhaps tongue in cheek – flirtation with scientism in Chapter 10 notes the selective advantages of a mathematical mind.

> "In a move that is now obvious, the Darwinian assumes simply that the rules of mathematics and logic, the basic beliefs about causality and the like, the epistemic values or principles [...] are at some level ingrained in our biology [...] One thinks mathematically because one is biologically disposed to do so, and one is attracted to simple and elegant theories for the same reason" (Ruse 2006, 242).

1.5. Biological science and the scientist's perception of what science is

Although John Cobb may be correct to infer that most practitioners "restrict themselves to a methodology that limits (biological) science to the study of what is purely objective" he is referring not to the science, but to the scientist's (technician's) perception of what science is. Subordination to techniques, reliance on "model" organisms and/or obedience to the received view that this is the only way to proceed, have confined the imagination. But biology as *science*[9] is not so restricted.

---

[9] A recent editorial of the online journal *PLoS Biology* eloquently describes the situation in North American universities: "Not so long ago, virtually every major university had a department of biology, or perhaps bookend departments of zoology and botany, which complemented physics, chemistry, mathematics, and possibly geology to form its science foundation. Biology was, at least compared to the field today, an integrated discipline, from the molecular and cellular to the ecosystem, firmly resting on Darwinian principles. Weekly colloquia drew biologists from across the spectrum, whether the topic was the genetic code, the nature of the synapse, or the Cambrian Radiation. But biology has seen its own radiation and is just starting to catch up with this explo-

Let us see what Whitehead goes on to say:

> "[...] The trouble with creators of today is that they try to substitute a mental idea for the aesthetic experience. They think: 'Look here, wouldn't it be exciting to try it this way: a way no one else has ever tried it before?' But the novelty is of no significance. All that has any significance is the depth and validity of an experience out of which the art comes; and if it comes out of mere consciously clever ratiocination, it is foredoomed" (Price 1954, 67).

I belong to a tradition in which Whitehead's remark is as true of science as of art. Creators of biological knowledge are *not limited* to what can be objectively measured (by us technicians) in line with objective criteria (incorrectly assumed to be) suitable to the requirements of the mathematical and physical sciences. The *non-technical ingredient* out of which the science comes is the depth and validity of experience. All living entities, all living relationships, analysed by objective methods received their *names* and the *attention* devoted to them as a result of subjective processes. My commitment to naturalistic (rather than strictly materialist) explanation includes both the subjective and the intersubjective in the scientific process.

1.6. The positive side

While I realize that this is not what most Whiteheadians have in mind, the following will illustrate something of what is meant here.

- *The Origin of Species* presenting Darwin's evidence for natural selection is a work of literature: nonetheless science. The tradition continued

---

sion. The amazing pace of advance in our understanding of biology has, perhaps unavoidably, engendered increasing specialization. Much of that advancement has involved the development of new tools, both in the laboratory and in computer models, and this has been dependent on the migration into biology departments of tools and people from physics, mathematics, chemistry, and elsewhere. These new collaborators have catalyzed rapid progress on specific problems, but they often have little interest in the broader scope of biology. Even traditional biologists with broader interests may not have the time to indulge outside of their own research areas because of the speed of scientific progress in those areas and the competitive nature of contemporary science" (Levin 2006, 300).

through Darwinism, the modern synthesis and Neo-Darwinism right up to the present day. Literary formulations of evolutionary theory – relying on metaphor – may fail as spectacularly as technical formulations, but they do not have the technical constraints of the latter (such as how many variables can be considered).

- Darwin's ideas – as he continually testifies – were born of subjective experience. He was above all a relentless observer. Theory was forced on him by observation and in turn forced him into further observation and experiment. The imaginative process goes all the way back to the infant's need to *make sense of the world* (Donaldson 1978). In Richard Gregory's definition of perceptions as hypothesis formation (Gregory 1966, 1980, 1997, 1998), it is difficult to say which comes first during the fact-forming process: the realities perceived, or the brain's value-based proclivity to perceive in certain ways. Creators of biological knowledge make great use of the comparative method (see below). Choosing what to study and what to compare are subjective processes. As to *intersubjectivity*, the facts of mimicry amongst butterflies in the South American jungle recorded by Fritz Mueller in the 19th century and Konrad Lorenz's observations that so advanced the science of ethology in the 20th, were the fruits of experiences shared with subjects in the insect and the avian[10] and canine worlds.
- Two 21st century contributions to biological theory will illustrate that Darwin's reciprocal procedure of passionately assembling experimental facts about "organisms" while continually meditating on their relation to the whole is still alive and well in some of today's scientists. The first overthrows a 100-year old theory about the transport of water in tall trees (Zimmermann et al. 2004). (Though having little to do with evolutionary epistemology, it is an instructive example of how long [largely because it was simple) a bad theory – the Cohesion Tension theory, barely supported by bad measurements – can be uncritically "accepted" by the biological community at large and taught in schools). The second is of an integrating principle operating amongst arrays of large molecules at the cellular and subcellular levels. *Conformational spread* (Bray and Duke 2004) is an intuitive leap that has enormous po-

[10] Lorenz's 1935 classic is entitled "The companion in the world of birds".

tential for the better understanding of self-organization, both generally and in evolution.

Finally, here is a less earth-shaking recent publication from the online journal *PLoS* (*Public Library of Science*) *Biology* – on cooperative hunting between fish of two different species (Bshary et al. 2006) – which treats of intersubjectivity in the underwater world. (In line with ruling paradigms the title of this work is neutral; the authors at several points apply the null hypothesis to their "purely objective" findings. Nevertheless, to have recognized in the first place that what they were seeing was cooperative behaviour is – while not stated – like Mueller's and Lorenz's an intersubjective experience).

1.7. Two populations of biologists: two biologies

It will be seen from the above that I am trying to *deflect* the choice that Cobb offers biologists.[11] Not because I do not think that biologists need to put their own house in order – the societal need is desperate[12] – but to provide a perspective that extends to the Enlightenment and beyond.

Cobb seems to be suggesting that the choice is between science and something to be added to science. I prefer to see the alternatives as a choice between the science of those professional biologists who already include "subjective aspects of nature" and "purposive character of living things in their explanations", and a reduced science of those who do not. Historically, biology is more the former than the latter (and lesser). When he writes: "Mainstream biologists know that there is evidence for all of the

[11] Cobb says: "Honest scientists, in my view, have two choices. They can continue to restrict themselves to a methodology that limits science to the study of what is purely objective and then recognize that there are other factors at work that they can not include. This means that they will acknowledge that scientific explanations are inherently incomplete. Or they can expand the study of science to include subjective aspects of the nature they study. They will then be able to include the purposive character of living things in their explanations".

[12] Unlike the medical profession, the professional bodies of most biologists do not have deontological rules regulating our relationships with the objects of study or that regard them as subjects. Such rules as exist, are imposed upon the profession from outside.

relationships they omit from their formulations of the mechanics of evolution" he is really referring to two populations of individuals – one making the science the other making certain "standard" formulations.

1.8. Truth is indivisible

I personally do not feel a need to subscribe to the full Whiteheadian explanatory principle of all nature as "occasions of felt experience" (Delafield-Butt 2007). In what amounts to a second essay, I shall nevertheless now try to show from my own experience how questions of evolution amongst the "higher" animals (as well as some traditionally "lower" creatures) can be approached by biologists with a metaphysical commitment to naturalism and without resort to dualism. Many biologists of similar tradition could no doubt tell a similar story of their careers, into which the "subjective" would always intrude as essential partner in the process of creating knowledge.

2. A different way of thinking

> "Biological facts, from genetics to neuropsychology, show that living systems can properly be said to act in pursuit of certain aims. Each tries to achieve certain standards appropriate to its way of life. The result of this continual striving, choosing and deciding, through millions of years, has been a progressive accumulation of information about how best to live. Contrary to what is often said the facts of biology show both purpose and progress in life" (Young 1978).
> "Biologists know that animals and even plants are not puppets, manipulated by the environment. They are *agents*, provided with targets and a remarkably strong inner motor tendency that causes them continually to strive to achieve the aim of remaining alive" (Young 1987, 7-8).[13]
> "It is the essence of life that it exists for its own sake, as the intrinsic reaping of value" (Whitehead 1934).

[13] This quotation is from the 1982 Shearman Memorial lecture series. Its original title, *Philosophers Use your Brains!*, was born of Young's experience that philosophers of mind did not take into account the exciting advances in anatomical and physiological knowledge of brain functioning marking much of his productive life.

In this second section I describe a way of thinking about the activities of animals – i.e. their physiology and behaviour operating through processes of pattern recognition – which grounds explanations of their evolution in principles applying to self-evaluating wholes.

2.1. The power of the comparative approach

Even though he claimed not to understand the philosopher's writings, it is evident from the two quotations above that J.Z. Young (1976) was something of a Whiteheadian. The background to the thinking is described by Peter Medawar (pupil of Young and eight years his junior). "Biology's central discipline [...] comparative anatomy has many of the virtues traditionally associated with the classical education called for by the humanities" (Medawar and Medawar 1983, 60). Its chief aim is the tracing of *homologies* – which embody both persistence and change during evolution. When linked to a pre-university schooling[14] in the humanities, we have a powerful combination. A contemporary influence was the Oxford "ordinary language" philosopher Gilbert Ryle. Ryle disdained mind-body dualism, putting it down to muddled thinking.

The octopus story. Young was very successful in studying the brains of octopuses and squids (cephalopods). These animals are almost as far genetically from our own line of descent as one can get, and therefore of special value for comparisons.

When I joined the small band of biologists and psychologists investigating their perception and learning abilities in Naples, I was able to point to the part they had probably played in their evolution from simple mollusk. The story is straightforward. When I look at an octopus which is looking at me[15],

[14] The home page of Marlborough College, which both Young and Medawar attended as schoolboys, describes it as "a community where scholarship is cherished, creativity is celebrated, diversity is evidenced, and conversation – the means by which knowledge is elevated into wisdom – is paramount".

[15] Much of my (also of Young's) understanding of biology was acquired from working on the octopus. I think it is not distorting the evidence to state that a critical part of that understanding came from establishing an *intersubjective relationship* with the animal.

I see that its eye is like mine. To find out what kind of scientific validity can be placed on this notorious instance of evolutionary convergence (or parallelism), I chose (subjectively) to measure (objectively) some parameters relating to functioning of eye and brain. I find a fish-eye lens, working on principles identical with those of its underwater compeers, feeding to an enormous brain whose circuit diagrams recall my own. Interest is quickened by finding that the sophisticated defense and camouflage repertoire of this dynamic and resourceful creature – epitome of *Metis* for the Greeks[16] – can regularly outclass my own search-and-find abilities.

Fortunately cephalopods have a long fossil history. Thanks to the good preservation of their shells, one can follow changes of behaviour over hundreds of millions of years every bit as radical as those traceable in the vertebrate skeleton.[17] The morphological parallels in the end-points of the two lines of descent were traditionally explained as the result of similar selection pressures from a common physical environment. As this explanation did not take functions into account, it did not convince me any more than it convinces Cobb who wishes to see a role for the activities of the phenotype in natural selection. I checked for records of predator/prey relationships between individual cephalopods and individual vertebrates now and in the past – and of other processes well-known to influence survival in a hostile world – and concluded that the remarkable parallels in dynamic life style were the result of competitive interactions in a "behaviour space" dominated by vision and centred upon food (Packard 1972). Convergent evolution (between cephalopods and vertebrates) is evolution of behaviour and is driven by behaviour. Metaphorically, one eye had "produced" the other.

2.2. *Emotions: their "luring or driving value" – a neuro-ethologist's synthesis*

To relate this to an evolutionary epistemology that takes us away from the collapsed (Ayala) explanation with its ratiocination about genes and point

(N.B. We are talking here about a relationship, not with a dog or a cat or other mammal – which even hard-headed scientists would readily concede have feelings – but with a mollusk!)

[16] *Metis* was the first wife of Zeus and goddess of ingenuity.

[17] Coupled with conservatism in the tissues of which the two are built.

mutations, without at the same time losing sight of the individual, I choose Erkki Haukioja's *Theory of Living Entities* (1982) which emphasizes the process of living (POL).[18] Haukioja's "*criterion for success of a living entity (automaton) is its functioning at the moment of evaluation*" rather than hypothetical future reproductive "success". It introduces self-evaluation as explanatory principle.

In her 2-volume work *Mind: an essay in human feeling* Suzanne Langer explores the role that feelings play in animal behaviour.

> "The basic assumption here proposed is the constant guidance of overt animal action by feeling [...] This bars any explanation in terms of social "usefulness" or prevision of future conditions. All the conditions are 'now' and the guidance, from the total impulse to consummation of the fully elaborated act, is by the agent's own feeling. The motivation of a behavioral act has to be conceived as a felt element in the situation from which it arises, that is, as something with a luring or driving value for the performing organism, not only as an inherited reaction established by 'natural selection' for the good of the species" (1972, 141).

For Whitehead (1934) the emotions have a time-transcending vector relationship to the process of living: they both proceed from and they aim towards.

It would have been impossible for Charles Darwin to have introduced the "felt element" into his mechanisms of evolution; his task was to explain the origin of species. But he did as much as any scientist before or since to

[18] POL assumes that life is a continuing of functioning which can not be defined by looking at structures. The criterion for success of a living entity (automaton) is its functioning at the moment of evaluation. On Haukioja's definition operational information (OI) for running automata can be divided into two logically different compartments: maintenance information (MI) and reproductive information (RI). By information is meant not the code, but the instructions produced when the code is translated. The emphasis is plainly on context and present action; continuity is important, not how many survive. This ecologist's experience of the evolution of life-history strategies had taught him that the genotype can "foster" the tactics and traits of any of a number of phenotypes "depending on developmental pathway and variable environment" (Haukioja and Jokela 2000, 180). (For the concepts of traits, tactics and strategies in Life History theory see Southwood 1977). (According to several Web based citation indices this carefully argued evolutionary theory (Haukioja 1982) has been cited less than 10 times in the past 30 years).

establish the primacy of the emotions in organic existence, first by demonstrating their commonality – his love of dogs (creatures originating a hundred million years apart from us) served him well in this; second by equating action with feeling;[19] third in being free from hang-ups about their place in science.

Without trespassing on the philosopher's sacred domain of qualia, it is nowadays much easier for the biologist to bring these thoughts into line with the facts. The analytical protocol adopted belongs equally with the animal psychologist,[20] ecologist, physiologist and comparative morphologist – Darwin was all four.[21]

The various forms that behaviour takes introduce directed changes into the ecosystems of which they are a part and alter their information content – the directional element being present in pattern recognising processes from the beginning of life. Evolution harnesses these changes, in turn altering the processes and enabling the evolutionary machinery itself to evolve. Special interest focuses on predators as selective agents and on the expanding power of the brain.

First predation. Predation is the most dramatic way in which one living entity can affect the survival of another[22] – simply by terminating the proc-

---

[19] In the Introduction to *Expressions of the Emotions in Man and Animals* written many years after publication of the *Origin of Species*, Darwin reflects on what one might call the "Darwin/Spencer law" that "feeling, [passing a certain pitch,] always vents itself in bodily action" (square brackets mine). "Man himself cannot express love and humility by external signs, so plainly as does a dog, when with drooping ears, hanging lips, flexuous body, and wagging tail, he meets his beloved master" (Darwin 1872, 12).

[20] In ethology, the recognized terms are motivation and drive (*Stimmung*).

[21] Morphologist and physiologist as author of *A Monograph on the Sub-Class Cirripedia* (1851-54), *The Movements and Habits of Climbing Plants* (1875), and *The Various Contrivances by which Orchids are Fertilized by Insects* (1877); one of the founders of ethology as author of *The Expression of the Emotions in Man and Animals* (1872), and of ecology with *The Formation of Vegetable Mould, through the Action of Worms, with Observations on their Habits* (1881).

[22] Cobb writes: "This is so obvious that it seems unnecessary to illustrate it. If a new predator enters an ecosystem, it certainly changes the environment for its prey. Habits that may previously have been adaptive may no longer be so".

ess of living. It is never a random process (see Curio 1976). The recognition processes which enable a prey item to be singled out from a population of potential prey range from simple and specific "innate releasing mechanisms" to complex interactions whose outcome depends heavily on the individual experience both of the predator (or predators) and of the prey. Such acts ("selection by the phenotype") are guided by feeling in the higher animals and are quite different from selection by physical characteristics of the environment.

The critical "survival events" (Packard 1988) may be very brief. The visual strike of a young herring at its first food item conditions all future strikes.[23] Ethologists speak of a "search image". Octopuses attack and eat crabs recognising them by sight, but only if the crab is moving. Not surprisingly, crabs of the genus *Carcinus* keep very still, or move very slowly once they have been observed. They defend themselves with their claws: holding them outstretched as warning display; if attacked, they use them to deliver a nip *painful* enough to cause a small soft-bodied octopus to withdraw (Packard 1988)[24]. The claws of the crab are here acting as *teaching* aids in a one-trial learning session that inhibits further attacks for a while – a result that momentarily benefits other members of the local crab population.

[23] Rosenthal (1969). Larval herrings, at hatching, are programmed to swim straight ahead; activity is conditioned by the "motivational" state of the fish (once the yolk supplied in the egg has been exhausted) and by olfactory cues; if potential prey is encountered a larva swims more slowly; the strike itself is preceded by a visual fixation lasting about 1 sec, with the body in a tight S-shape; this is the critical moment when an individual prey item is identified and selected or rejected and spared. Some form of imprinting seems to be at work; in experimental conditions, individual larvae tend to continue feeding on the prey species of their first successful encounter.

[24] The claws of the crab *Carcinus mediterraneus* (species commonly preyed on by *Octopus vulgaris*) are not adapted to pierce or to sever; the tips are blunt and produce localized pressure through the sustained contraction of large specialised "catch" muscles specifically adapted to evoke *nociception*. (The pain – experienced by anyone prepared to be nipped by a crab – has a special quality associated with capsacin). Immobility, "claws display", "catch" muscles, blunt claws for delivering pain, are all "coadapted *traits* that together make a *tactic*". In the innate defence strategy of *Carcinus*, the tactic is to promote learning through punishment of the predator.

During the trial and error experiments for training octopuses in the laboratory (see Young 1961, Sanders 1975), the negative reinforcement in such "teaching" is an electrical shock. The positive reinforcement (or "reward") signal is the *taste* of food: in both cases, a *feeling.* These things are encoded in their brains.

The brain as symbol of emergent evolution. The behaviour of an automaton – considered as that part of its activities in communication with the outside world – can be crudely represented by the dimensions of its growing brain: or, more meaningfully, by *brain/body weight proportion.* In some lines of animals (e.g. crabs) brain/body weight ratios have changed little over the last 300 million years; in others they have changed at accelerating rates always associated with increased capacity to deal with the environment and to learn. For Allan Wilson, the ascending curve – proceeding at a rate greater than exponential – is itself evidence of an autocatalytic process. He coined the terms "pressure to evolve" and "cultural drive" to encompass the idea that "the brain of mammals and birds is the major driving force behind their organismal evolution" (Wilson 1985).[25] When I came to look at the brain/body weight ratios of various cephalopods (squids, cuttlefish and octopuses) I found the log-log curves describing them (Packard 1972) to lie amongst those of higher fishes. The operational information (OI) acquired by an octopus during its lifetime takes up space. Genome size is little or not affected.

[25] Alan C. Wilson (1985) plotted relative brain sizes of vertebrates (Jerison's data) against the times of origin of the tetrapod groups on the palaeontological scale. The resulting curve gives an accelerated rate of change amongst higher vertebrates even faster than that for other morphological changes plotted in the same way. The finding is particularly dramatic coming from a molecular biologist since the rates of molecular change over the same period, whether measured as mutation rates (rates of substitution of base pairs in DNA) or as rates of change in other macromolecules, has altered no more than two- or three-fold. (However, if any term from physics is to be employed to describe the brain's consequencies for evolution it should perhaps be "power" – the rate of doing work – rather than a force such as "pressure"). Wilson proposed a mechanism by which one particular form of learning – imitation by conspecifics – can speed up the evolutionary process, by altering the rate at which mutations become fixed. A trait which spreads laterally through a population by being imitated will produce conditions in which mutations favouring or improving that trait will be more likely to be fixed.

Perception as evaluation. Ensembles of classifying cells of the deep retina situated on the first step of the visual pathway – whether of my eye or the octopus's – report the biologically relevant abstract properties of the visual world: appearance/disappearance, contrast, relative movement, etc. They do not signal absolute levels of illumination (see Figure 1) or illuminating wavelength – or mass displacements of the visual field occurring because my head turns.

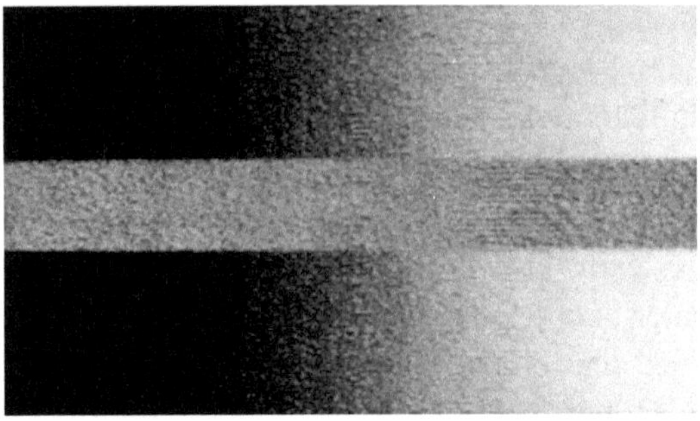

Fig. 1: *Simultaneous contrast illusion.* It is a measure of the generality of the information-processing involved in the illusion, that the neurophysiological basis of the process – lateral inhibition – was first discovered in the compound eye of an invertebrate animal: the horseshoe crab *Limulus* (Hartline et al. 1956). In humans, also, it takes place within the eye (see Asher 1951) and cannot be consciously suppressed. To perceive the central strip as of one mean grey level, it is necessary to mask the flanking strips; the combination of luminance and texture in this illustration seem to enhance its effect. (Source: Frisby, J. (1979).*Seeing: Illusion, Brain and Mind.* Oxford: Oxford University Press.)

Each subsequent step on the visual pathway performs further abstractions. Ballard, following Barlow (1972), referred to visual cortical cells as "value units" (see Ballard 1986 and associated commentaries). The extent to which each is tuned to (i.e. fits) derivatives of the scene encoded in their firing frequency, is a quantity that can be measured. Equally important is the finding that such "fit" is the result of feedback from the environment during development; the process in "higher" animals takes many months or years. Perhaps physiologists would do better to write of "meaning-making" rather than "information-processing".

In a brilliant investigation of the tuning of binocular cells of a kitten during POL – considered as a model for learning at the cellular level – Wolfgang Singer (1987) placed the process of evaluation actually inside the cell. The "reward" signal received by these units, generated by behaviour that has achieved its object, reaches the cortex by way of ascending pathways from brainstem areas close to neurochemical centres associated with mood and emotions.

Habitat selection by sessile marine organisms – such as bryozoa, tubeworms, barnacles, mussels, oysters, and many coelenterates – is a prime example of a means by which living entities assess their fitness to survive in a particular location of *proven value* (Buss 1979). For most of their lifetime they are found encrusting rocky shores above and below the tide line and have little or no power of independent movement. Many are gregarious. The swimming larvae (or dispersal phase) of such creatures actively choose the spot on which to spend the rest of their lives. After a period in the plankton, if appropriate conditions are not encountered when they are ready to settle they can prolong the search for days or weeks. A barnacle cyprid larva is equipped with a barrage of sense organs and responses tuned to this end; it first searches widely, then closely and then inspects the site before cementing itself into a place to metamorphose and grow up – amongst other barnacles (Crisp 1976).[26]

The path of a salmon off the east coast of Scotland seeking to spawn in the stream in which it was born is an enlarged version of that taken by the spat of a barnacle seeking a specific site on which to settle. The address of the home stream is imprinted in salmon as smolts and can be recorded electrophysiologically either at the single neurone level or at the level of the enormously increased EEG (electroencephalogram) and swimming activity it triggers. Seawater not smelling of the home stream is ignored and the fish proceeds on up the coast until it encounters water that does. Once it has entered the river and moves upstream, the selection process reiterates as it samples its tributaries. The interest of this example is that the mechanism enables a salmon to recognize a habitat (or refuge) of proven worth

[26] The adhesive organ of cyprid larvae is able to recognise the surface properties of the tanned protein arthropodin present in the cuticle of other barnacles (Crisp 1976).

(value) for its offspring through direct reference to the encoded experience of the parent when a juvenile: fitness being inferred from survival of the fish to date.[27] Though the accompanying feelings can not be recorded, at least the drive component can.

All biology is pattern recognition of one kind or another (Roederer 1979). Cell biologists, investigating the sorts of *relationships* that John Cobb says are missing from Ayala's account of the evolutionary mechanism, explicitly acknowledge this in their everyday use of the words "recognize" and "recognition" to describe interactions taking place between macromolecules at the cell membrane.

Towards the other end of the spectrum of relationships characterizing the living world – in the part called *sociality* – subjects (or automata) are continuously recognizing the evaluation process and its conveyed intentions in the actions and expressions of other subjects (for discussion see Emery and Clayton 2009). One day, the role that feelings play in this implicit *understanding* between subjects and their contribution to the history of the biosphere will doubtless find a formal place in evolutionary theory.[28] [29]

---

[27] Buss (1979) has examined the relation of such events to the concept of "Neo-Darwinian fitness" and introduced the name *refuge* for those sites that fulfil the requirements for future survival and growth of the adult. He defines refuges as "spatial positions on marine hard substrata where the fitness of an individual is high relative to other spatial positions". Since larvae are actively recognizing refuges, this definition corresponds with Haukioja's criterion of fitness.

[28] Cobb writes: "For Whitehead subjective experience and what is perceived objectively belong together in a single world. The actual course of evolution is a product of both working together". Here I am placing emphasis on the experience of subjects in the living world, rather than of the scientist whose subjective experience of that world is engaged in studying it (see above).

[29] The above examples of behavioural processes which recognize and store the memory of causal relations in the environment, are taken from an unpublished essay written more than twenty years ago ("The Role of Behaviour in Evolution: a Metaselection Theory Rooted in Pattern Recognition" – running title "Evolution by Metaselection" (available on request)). Instances of the "Baldwin effect", it is the details that count. Metaselection theory has still to be axiomatised. It leads ultimately to the abstraction that behaviour is the stable bloc upon which the genome is slipping. Understanding of motives and intentions provides a teleonomic machinery (*metaselection*) for "innovative" evolution which is recognising its own processes in other living entities. (The

3. Summary and conclusions

I have tried to deflect John B. Cobb's criticism of the simplified (Anglo-American) version of Neo-Darwinism by pointing out that most of those who "accept" it are not required to test it. Its place in current teaching reflects cultural expectations and realities: reliance on technical solutions to complex problems, the fashion for emphasis on the competitive in daily life to the detriment of the associative, the reductive "numbers game" in economics, etc.

Second, while evolutionary theory is expanding just as rapidly as other areas of thought, all Neo-Darwinian formulations predicting the fate of genes or populations have a future reference: uncoupled, therefore, from the work of "most biologists" concerned with living processes in the present – or the story of evolution in the past.

Third, there is a long tradition of Darwinian biologists who attribute to behaviour and the activities of the phenotype an important role in directing the course of evolution. These can include "subjective aspects of the nature they study".

The second half of the chapter develops this Whiteheadian message by drawing on my own experience of psycho-physics and of the forms taken by pattern-recognition in the life histories of aquatic organisms. It incorporates the ecologist Erkki Haukioja's little known Theory of Living Entities which emphasizes the process of living (POL) and places self-evaluation at the heart of evolutionary dynamics.

References

Allee, W. (1938). *The social life of animals*. London and Toronto: Heinemann.

Asher, H. (1951). "Contrast in eye and brain". In: *British Journal of Psychology*, 40, pp. 187-194.

Ballard, D. (1986). "Cortical connections and parallel processing: structure and function". In: *Behavioral and Brain Sciences*, 9, pp. 67-120.

---

word "metaselection" was coined independently in an unpublished essay by Tony Smith (1993) http://www.meme.com.au/theoria/metaselection.html)

Barlow, H. (1972). "Single units and sensation: a neuron doctrine for perceptual psychology?" In: *Perception*, 1, pp. 371-394.
Bateson, P. (2004). "The active role of behaviour in evolution" (Review of Weber, B.; Depew, D. (2003). *Evolution and Learning: the Baldwin Effect Reconsidered.* MIT Press). In: *Biology and Philosophy*, 19, pp. 283-298.
Bray, D.; Duke, T. (2004). "Conformational Spread: the propagation of allosteric states in multiprotein complexes". In: *Annual Review of Biophysics and Biomolecular Structure*, 33, pp. 53-73.
Bshary, R.; Hohner, A.; Ait-el-Djoudi, K.; Fricke, H. (2006). "Interspecific Communicative and Coordinated Hunting between Groupers and Giant Moray Eels in the Red Sea". In: *PLoS Biology*, 4, No. 12, p. 431.
Buss, L. (1979). "Habitat selection, directional growth and spatial refuges: why colonial animals have more hiding places". In: Larwood, G.; Rosen, B. (eds.). *Biology and systematics of colonial organisms*. London, New York, San Francisco: Academic Press, pp. 459-497.
Cooper, G. (2003). *The Science of the Struggle for Existence: on the Foundations of Ecology.* Cambridge: University Press.
—— (1988). "Fitness and Explanation", in: *PSA*, 1, pp. 207-215.
Crisp, D. (1976). "Settlement responses in marine organisms". In: Newell, R. (ed.). *Adaptations to environment: essays on the physiology of marine organisms*. London: Butterworths, pp. 83-124.
Curio, E. (1976). *The ethology of predation*. Berlin, Heidelberg, New York: Springer.
Darwin, C. (1904). *Expression of the emotions in man and animals*. London: Murray, 1872; 2nd (popular) edition.
Delafield-Butt, J. (2007). "Biology". In: Weber, M.; Seibt, J.; Rescher, N. (eds.). *Handbook of Whiteheadian Process Thought*, Vol. II. Heusenstamm: Ontos, pp. 157-168.
Donaldson, M. (1978). *Children's Minds*. Glascow: Fontana/Collins.
Dover, G. (2000). *Dear Mr. Darwin: Letters on the Evolution of Life and Human Nature*. London: Weidenfeld and Nicolson.
Emery, N.; Clayton, N. (2009). "Comparative Social Cognition". In: *Annual Review of Psychology*, 60, pp. 87-113.
Frisby, J. (1979). *Seeing: Illusion, Brain and Mind.* Oxford: Oxford University Press.
Grafen, A.; Ridley, M. (eds.) (2006). *Richard Dawkins: how a scientist changed the way we think.* Oxford, New York: Oxford University Press.
Gregory, R. (1980). "Perceptions as Hypotheses". In: *Philosophical Transactions of the Royal Society, London* B 290, pp. 181-197.
—— (1966). *Eye and Brain: The Psychology of Seeing*. London: Weidenfeld and Nicolson, Fifth Edition (1997) Oxford University Press and (1998) Princeton University Press.

Hartline, H. ; Wagner, H.; Ratliff, F. (1956). "Inhibition in the eye of Limulus". In: *Journal of General Physiology*, 39, pp. 651-673.
Haukioja, E. (1982). "Are individuals really subordinated to their genes? A theory of living entities". In: *Journal of Theoretical Biology*, 99, pp. 357-375.
Haukioja, E.; Jokela, J. (2000). "Evolution of strategies to stay in the game". In: *Biology and Philosophy*, 15, pp. 177-196.
Koutroufinis, S. (this volume).
Langer, S. (1972). *Mind: an essay in human feeling*. Baltimore: John Hopkins Press.
Levin, S. (2006). "Fundamental Questions in Biology". In: *PLoS Biol*, 4 (9), p. 300.
Lorenz, K. (1935). "Der Kumpan in der Umwelt des Vogels". In: *Journal für Ornithologie*, 83, pp. 137-213; 289-413.
Medawar, P.; Medawar, J. (1983). *Aristotle to Zoos*. Cambridge, MA: Harvard University Press.
Molyneaux, P. (2007). *Swimming in Circles*. New York: Thunder's Mouth Press.
Packard, A. (2006). "Contribution to the whole (H). Can squids show us anything that we did not know already?" In: *Biology and Philosophy*, 21, pp. 189-211.
—— (1998). "Visual tactics and evolutionary strategies". In: Wiedmann, J.; Kullmann, J. (eds.). *Cephalopods: present and past*. Stuttgart: Schweitzerbart'sche Verlagsbuchhandlung, pp. 89-103.
—— (1972). "Cephalopods and fish: the limits of convergence". In: *Biological Reviews*, 47, pp. 241-307.
Price, L. (1954). *Dialogues of Alfred North Whitehead*. London: Max Reinhardt.
Reid, R. (2007). *Biological Emergences: evolution by natural experiment*. Cambridge, MA; London: MIT Press.
Roederer, J. (1979). "Human brain functions and the foundations of science". In: *Endeavour*, 3 (3), pp. 99-103.
Rosenthal, H. (1969). "Untersuchungen über das Beutefangverhalten bei Larven des Herings *Clupea harengus*". In: *Marine Biology*, 3, pp. 208-221.
Ruse, M. (2006). *Darwinism and its Discontents*. Cambridge: University Press.
Sanders, G. (1975). "The cephalopods". In: Corning, W. ; Dyal, J.; Willows, A. (eds.). *Invertebrate Learning*, Volume 3. New York, Plenum Press, pp. 1-101.
Singer, W. (1987). "Activity dependent self-organisation of synaptic connections as a substrate of learning". In: Changeux, J-P.; Konishi, M. (eds.). *The neural and molecular bases of learning*. New York: Wiley 1987, pp. 301-336.
Southwood, T. (1977). "Habitat, the templet for ecological strategies?" In: *Journal of Animal Ecology*, 46, pp. 337-365.
Whitehead, A. (1934). *Nature and Life*. Chicago: University of Chicago Press.
Wilson, A. (1985). "The molecular basis of evolution". In: *Scientific American*, 253, pp. 148-157.
Wingreen, N.; Levin, S. (2006). "Cooperation among Microorganisms". In: *PLoS Biology*, 4 (9), pp. 1486-1488.

Young, J. (1987). *Philosophy and the brain*. Oxford: Oxford University Press.
—— (1978). *Programs of the brain*. Oxford: Oxford University Press.
—— (1976). "Choice, determination and value in the light of biological knowledge". In: *Journal of Theoretical Biology*, 62, pp. 459-465.
—— (1961). "Learning and discrimination in the octopus". In: *Biological Reviews*, 36, pp. 32-96.
Zimmermann, U.; Schneider, H.; Wegner, L.; Haase, A. (2004). "Water ascent in tall trees: does evolution of land plants rely on a highly metastable state?" In: *New Phytologist*, 162 (3), pp. 575-615.

Response to Andrew Packard

JOHN B. COBB, JR.

I am deeply appreciative of Dr. Packard's response to my essay. I interpret it to express full agreement with me that the neglect of the activity of the phenotype in evolutionary theory is a mistake. Indeed, he has documented the role of this activity richly in many different contexts.

Further, he agrees with me that when we consider the activity of animals we cannot avoid considering their subjective side. He recognizes that animals have sensations and emotions and purposes. He sees no reason to exclude these from the account of animal activity. He reminds his readers that there is a long tradition of Darwinian evolutionists who have paid attention to animal activity and have not excluded the subjective dimension of animal experience. I named a book I edited recently Back to Darwin in order to emphasize that Darwin was open to much richer paths than the one taken by Neo-Darwinians.

So far as I can tell, Packard agrees with me also that Neo-Darwinists do neglect the activity of animals and systematically exclude consideration of their subjective experience. So I assume he agrees that Neo-Darwinism is inadequate by virtue of its omission of what I called "the fourth variable".

I am truly pleased by these extensive agreements. Nevertheless, Packard thinks of his paper as deflecting my criticism. I think one of his points is that I imply that the mainstream of evolutionary biology is committed to the Neo-Darwinism I criticize. He thinks I am wrong in this, that even though many biologists give lip service to the Neo-Darwinian formulation, their actual work ignores it and takes much into account that Neo-Darwinism ignores. I hope he is correct.

Nevertheless, I do not consider it a healthy situation when the textbooks that introduce students to biology affirm an inadequate theory that most practicing biologists ignore. That theory is, unfortunately, not ignored by high school teachers of biology, and it is a cause of quite serious, social friction, which, if Packard is correct, is quite unnecessary. It would be a great benefit to society if those numerous biologists who do not in fact believe the Neo-Darwinian theory of evolution would speak up. Giving them voice as Packard does in this paper is a valuable contribution. But until they undertake to improve the textbooks and the public face of evolutionary biology, some of us outsiders will need to continue to criticize. Since there is very little activity in this direction among practicing biologists, it is hard for the outsider not to suppose that Neo-Darwinism is the dominant theory and that most biologists accept it. Perhaps we could have a campaign to have Erkki Haukioja's "Theory of Living Entities" taught in courses in biology.

Packard follows Whitehead a long way, but not all the way. His limits are probably of two sorts. Whitehead considers that every actual entity is, in its moment of existing, a subject, acted on by past entities and involved in its own self-constitution. Packard seems to limit subjectivity to living things. From a Whiteheadian perspective, this introduces the dualism that Packard rightly wants to avoid. But since our essays are about living things, I will not pursue this here.

Secondly, while Packard is open to some features of subjectivity, he may not be open to all. Donald Griffin, in The Question of Animal Awareness, drew up a list of terms dealing with subjectivity (58). It begins with those that are widely used by students of animal behavior and goes to those that are most avoided. It ends with "consciousness" that is, Griffin states, taboo. The most acceptable term is "pattern recognition". In between he includes terms such as affect, spontaneity, intention, feeling, and awareness, in that order of increasing reluctance. Griffin thinks that the working principle is to see how far one can go in explaining animal behavior without including subjectivity and then introduce only those that are amenable to research methods (55). Probably Packard would follow some such policy.

I suspect that, for many of those who allow some terms referring to subjective states but avoid others, the issue is how easily it may be supposed that these aspects of subjectivity are directly explicable physiologically. As long as they can be explained in this way, the subjective term can be regarded as shorthand designation of the physiological state that is the real cause. If this is the deeper explanation, the scientist has not really broken the bounds of what is regarded by many as an essential feature of science, that is, that physical things are explained by other physical things in a closed circle. By the "physical" here is meant that which is accessible, or thinkable, as object of experience rather than as subject.

In principle this would mean that a physicalist determinist, that is, one who understands all subjective states as epiphenomenal or supervenient on physical ones, could use any subjective terms with equal comfort. Packard's frequent identification of an animal with an automaton would fit this pattern, but I do not understand this reductionism to be his intention. If I am wrong, then, of course, the apparent similarity of our views would disappear.

Packard does not tell us explicitly the nature of the dualism he wishes to avoid. However, it is my assumption that it is one that allows a separate causal efficacy to subjective states. If the vast majority of science succeeds in finding causal explanations in the objects of experience, and then, in dealing with animals and human beings, we posit that the subjects of experience also exercise causal efficacy, that would indeed introduce a special kind of dualism.

Whitehead avoids that dualism in a different way. As David Hume showed long ago, and as strict empiricists have all agreed, there are no necessary relations of the causal sort between the objects of sense experience. One patch of color may be succeeded by another, but it does not cause the other. Of course, one can describe complex patterns that regularly recur, but regular recurrence says nothing about where causality is located.

Our certainty that there is real causality arises, not through these observations, but in personal experience. Strong pressure on my finger causes me pain. I experience my pain as caused by the pressure on my finger. The pain causes me to pull my hand away from the source of pressure. Or I may experience my anger as caused by the hurtful words of another. I may relieve the anger by lashing out at the one who caused it. The strong belief that we live in a causal world comes from experiences of this sort.

For Whitehead, all causality is the effect of some actualities on other actualities. The cause, in its moment of existence, is a subject receiving the causal influences of its past. As it completes its self-constitution it becomes an object for subsequent subjects. In this capacity it exercises its causal efficacy in their self-constitution. There is a vast variety of causes, but in this vision, there is no dualism of types of cause. There is, obviously, no exclusion of subjects from a role in this process.

There is a great deal in Packard's essay that breaks out of the narrow bounds of Neo-Darwinism and of materialistic science in general. Like Packard, Whiteheadians understand themselves to be naturalists but not materialists. We can read Packard's account with extensive agreement and great pleasure. Since, at the same time, Packard makes it clear that he does not want to go all the way with Whitehead, I have tried to guess what sorts of limits he posits and where he draws the line. If he really wants to avoid dualism, I would encourage him to remove the lines altogether. By removing the lines, Whitehead has overcome the dualisms Packard opposes. But I do realize that it may be too much to ask of a scientist to overcome fully the suspicion of the subjective sphere that has been endemic to science since it took its materialist turn in the seventeenth century.

References

Griffin, D. (1976). *The Question of Animal Awareness: Evolutionary Continuity of Mental Experience*. New York: Rockefeller University Press.

Erkki Haukioja to the Rescue?

ANDREW PACKARD

John Cobb's call for those practicing biologists who feel as I do, to club together "to improve the textbooks and the public face of evolutionary biology" does not fall on deaf ears. At a time when the ruling paradigms in so many arenas – political, financial, environmental – are being questioned as never before, such change would slip naturally in with the others that are hopefully to come. Visceral belief in selfish genes drew heavily on the language of market forces and the now discredited stock market – hence the title of this reply.

It will be tough going. Item: we are in 2008; a 2-day conference entitled "The Driving Forces of Evolution", held in the meetings room of the Linnean Society on the 150th anniversary of the reading of the famous Darwin-Wallace paper, did not include the role of Behaviour! My comment: "Driving forces without a driver!"

Though not explicitly touched on, my broad agreement with John Cobb goes beyond such sins of omission. I am not a theologian. The need for a formulation to replace ruling Neo-Darwinism derives from a "natural theology" that sees man's – and therefore also the biologist's – relationship with the universe as a moral one. In Naples some years ago, I organized a "Conversation" at the Italian Institute for Philosophical Studies to explore the abstract relationship between the Logic and the Ethic of the Life Sciences. We focused on a question put by Emilios Bouratinos: *can a science be considered ethical if it does not have a logic adequate to its subject matter?*

Within the framework of evolutionary theory – and, by extension, of the material products of evolution – the issue can be simply stated. *Ought* (Neo-Darwinian) genetic determinism to be governing our relationship with the biosphere? To take an extreme example: the physicist Freeman Dyson, fascinated by the potential "benefits" of biotechnology, suggests that one third of the earth's forest cover be bio-engineered in order to lock up carbon and so reverse climate change! Is this to be treated as a moral question, or only as an intellectual, technical and political one?

Erkki Haukioja asks his own intriguing question (whose inferred double negative implies interesting states of denial which we have no space to discuss). "Do we really have good reasons why the evolutionary theory is not based upon a sound theory of life and living organizations?" (Haukioja 1993, 14). Answer: "[…] we do not have in wide use any explicit theory of organizations which live". Without claiming to replace it, he invalidates the logic of the theory at fault: a) by supplying a (missing) definition of the living units in which evolution is seen to take place, b) by needing to see them as products of functioning, which are all the time being tested by the process of living (his POL, tantamount to Cobb's "fourth variable"), before any examination of how genetic dynamics may be ruling their functioning, c) by calling for an understanding, in cybernetic terms, of the general principles of living organization and the constraints they set upon evolution, and d) placing the replication of multicellular organisms (lineages) in a separate box from the rest of POL.

Haukioja's Theory of Living Entities (1982) has recently been incorporated into ecology (autecology) by Gimme Walter (2008). Both practical ecologists point out the persistent (and pernicious) blurring of two distinct things: the self-evaluating functioning of individuals and small populations on one hand, the details (such as genes) that are helping to achieve functioning on the other. Haukioja concludes that a "Neo-Darwinian theory (which) views everything via changes in gene frequency […] has made it unnecessary, perhaps even uninteresting, to analyze the logic of living" (Haukioja 1993, 24). Philosophically, we are confronted with the fallacy of misplaced concreteness, empirically, with research programmes so different that it is difficult to see where there can be conjuncture. One programme researches populations in which individuals become theoretical abstractions, the other individuals (whether unitary or modular) in which populations become abstractions.

On attitudes to consciousness raised by Cobb, I have two practical points to make. The present ascendancy of cognitive sciences has left few items on Griffin's thirty-years-ago list of taboos constraining behaviour studies. Moreover, the mainstream Neo-Darwinian theory under attack has not been built on data from the behavioural sciences. Admittedly, nor has

Haukioja's. (He only briefly alludes to subjective experience at the end of his 1993 paper – not to dismiss it, but to argue for his theory equally in the nuanced human context.) I too baulked at the word "automaton" when I first encountered it in Haukioja's writing. But for the purposes of his argument that the entities known to evolve are self-evaluating individuals (automata) using *information* for functioning – not "vehicles" for transporting genes in time – it was not relevant whether the information is built-in, acquired socially, subjectively, unconsciously, or consciously. None is excluded.

The second practical point obliquely touches on whether I personally am a dualist or an emergentist. Awed by the miracle that *anything* works, I find consciousness rather overrated! Consciousness fails, and errs during POL, in a way that unconsciousness does not. (Consider, moreover, the prospect that man, through the exercise of consciousness, is precipitating a collapse of the biosphere and the products of evolution to date, matching in its scale any of the catastrophic "events" of the past 500 million years.) Fortunately consciousness is held in check by the unconscious. Falling asleep is a beautiful example of Haukioja's cybernetic functioning. Tuned by POL in its myriad branches throughout that 500 million years, this universal (and stereotypical) adaptive mechanism does not fail during an individual's life.

Are the minds that have unravelled the multiple cause-effect relationships underpinning physiological sleep closer to Hume's framework of reasoning on causation or to Blaise Pascal's, I wonder?

Perhaps John Cobb – and the wider readership in this interesting exchange – will be disappointed that my response should be about technicalities. I hope not. I feel more at home in his company than I do with evolutionary theorists who chose to ignore Haukioja's ideas over the past quarter century, and than the metaphorical millions of colleagues who in their teaching of evolution keep in circulation a pre-emptive language and logic of "fitness" and "success".

Over to you, John.

References
(not included in original response)

Griffin, D. (1976). *The Question of Animal Awareness: Evolutionary Continuity of Mental Experience*. New York: Rockefeller University Press.

Haukioja, E. (1993). "What is the basic question in biology?" In: Kull, K.; Tiivel, T. (eds.) (1993). *Lectures in Theoretical Biology: the second stage.* Tallinn: Estonian Academy of Sciences, pp. 13-25.

Walter, G. (2008). "Individuals, populations and the balance of nature: the question of persistence in biology". In: *Biology and Philosophy*, 23, pp. 417-438.

Evolution without Tears: A Third Way beyond Neo-Darwinism and Intelligent Design[1]

DAVID R. GRIFFIN

A central dimension of the current cultural wars is the conflict between Neo-Darwinism and Intelligent Design. Many would call it simply a conflict between *evolution* and Intelligent Design, because belief in evolution has, by both advocates and critics, been widely equated with belief in Neo-Darwinism. Neither side, in any case, is going to convert the other, but a way forward might be provided by a third way in which the most vital concerns of each side would be reconciled. Such a position is needed because each side has valid criticisms of the other. Advocates of Intelligent Design, in charging that Neo-Darwinism is problematic from metaphysical, scientific, and religious-moral points of view, are not entirely wrong, and Neo-Darwinists, in charging that Intelligent Design is anti-scientific, are not entirely wrong. Insofar as these are the only two positions that are mentioned in public discussions of evolution, the public is forced to choose between two positions that are arguably about equally false.

The public discussion about the confrontation between these two positions has generally been unhelpful, largely because of failures to clear up ambiguities in some of the key terms, such as "evolution", "Neo-Darwinism", "naturalism", and "supernaturalism". I will show how a position based on the philosophy of Alfred North Whitehead can help us clear up these ambiguities and thereby point to a third way that could in principle

[1] A previous version of this chapter was presented in the session "Biophilosophy", organized by Spyridon Koutroufinis, as a part of the Sixth International Whitehead Conference, which took place in Salzburg, Austria in 2006. This chapter was previously published as the monograph *Evolution without Tears: A Third Way beyond Neo-Darwinism and Intelligent Design* from P & F Press (Claremont, CA) in 2006.
P & F Press has granted full permission to the editor of this volume to reprint this text.

be accepted by both the religious and scientific communities. This position will thereby demonstrate that evolution need not be a source of conflict between these two communities between evolution can be part of a worldview that is fully religious in a theistic sense while being fully scientific. In the final section of this essay, I will briefly discuss the implications of this position for the question of the teaching of evolution in public (state-supported) schools.

1. Neo-Darwinism

Much of the confusion surrounding debates about evolution revolves around failures to specify what one means when one is either defending or criticizing something called "Neo-Darwinism". Defenders tend to use the term interchangeably with "evolution", so that to reject Neo-Darwinism would be to reject evolution. They also regard Neo-Darwinism as scientifically proven, so that to reject it would be to reject science. From the perspective of Intelligent Design advocates, by contrast, Neo-Darwinism is less a scientific than a metaphysical doctrine, and a false one at that. And this metaphysical position, by portraying the universe as meaningless and devoid of moral norms, has pernicious consequences. Some defenders of Neo-Darwinism reply that it, as a purely scientific doctrine, is fully compatible with a religious-moral view of the universe.

These debates have been interminable partly because each claim is correct, given a particular conception of what Neo-Darwinism is. I will explain this point by laying out 13 doctrines constituting what can be called "maximal Neo-Darwinism". Although many advocates of Neo-Darwinism work with a more minimalist understanding, this maximal Neo-Darwinism, which is the position attacked by Intelligent Design advocates, is the position presented to the public by some of Neo-Darwinism's most influential advocates. The composite nature of Neo-Darwinism in this maximal sense is evident by the fact that these 13 doctrines need to be divided into three types: scientific, metaphysical, and religious-moral doctrines. The scientific doctrines, moreover, need to be divided into two levels: basic and derivative.

1.1. Two basic scientific doctrines

1. *Microevolution*: This doctrine affirms the occurrence of minor genetic and sometimes phenotypical changes within a species, or even the transformation of members of a species into a new species in one technical sense of that term. Sometimes called Darwin's *special* theory of evolution, this doctrine, which contradicts the idea that species are *absolutely* fixed, is now uncontroversial.

2. *Macroevolution*: This doctrine says that all present species (except perhaps the simplest organisms) have in some way descended from previous species over a long period of time. Darwin said that this doctrine – through which the idea of "descent with modification" replaced the idea that each species was separately created – was of more importance than his particular conception of how this descent occurred (Gillespie 1979, 130).

1.2. Five metaphysical doctrines

3. *Naturalism$_{ns}$:* Although the term "naturalism" is used for many different positions, there is a *generic* naturalism, which stipulates merely that there can be no supernatural interruptions of the world's normal causal processes. This generic naturalism can be called *naturalism$_{ns}$*, with "ns" standing for "nonsupernaturalist". Although it may seem redundant to speak of "nonsupernaturalist naturalism", this term distinguishes naturalism in this minimal, generic sense from a more restrictive doctrine, naturalism$_{sam}$, to be discussed below.

Naturalism$_{ns}$ is a metaphysical rather than a scientific doctrine, because it cannot be empirically verified. It is, nevertheless, part and parcel of "the scientific worldview", because it has become accepted as science's most basic philosophical presupposition. As philosopher of science Robert Pennock says, science needs "the binding assumption of uninterruptible natural law" (1999). Whitehead agreed, saying that "the full scientific mentality [...] instinctively holds that all things great and small are conceivable as exemplifications of general principles which reign throughout the natural order", so that "every detailed occurrence can be correlated with its antecedents in a perfectly definite manner, exemplifying general principles" (1967, 5, 12).

The doctrine of naturalism$_{ns}$ stipulates that all explanations of macroevolution must be entirely naturalistic in the sense of not implying any interruptions of, or insertions into, the world's normal causal processes.

4. *Uniformitarianism*: This doctrine stipulates that only causal factors operating in the present can be employed to explain past developments. In Darwin's own mind, this stipulation involved two dimensions: *ontological* uniformitarianism, which ruled out (among other things) supernatural divine interventions, and *geological* uniformitarianism, which ruled out occasional catastrophes. Today, geological uniformitarianism is no longer affirmed, but ontological uniformitarianism is absolutely presupposed.

5. *Positivism-Materialism*: Positivism (as used in discussions of evolutionary) is the doctrine that all causes of evolution must be at least potentially verifiable through sensory observation. This insistence is virtually identical with the insistence on exclusively material causes, because only such causes are in principle detectable through the physical senses. Positivism and materialism, accordingly, have the same implications, so we can combine them into one doctrine, *positivism-materialism*.

6. *Atheism*: Darwin himself was not an atheist, but endorsed what is now usually called *deism*, according to which God, after creating the world, no longer influenced it.[2] But Neo-Darwinism is fully atheistic, seeking to understand how our world could have arisen on the assumption that it is not the product of a divine creator.

In light of Neo-Darwinism's atheism and positivism-materialism, we can say that it embodies *naturalism*$_{sam}$, with "sam" standing for "sensationist-atheistic-materialistic". This doctrine includes naturalismns but goes

[2] Darwin's deism is illustrated by his statement that "some few organic beings were originally created, which were endowed with a high power of generation, & with the capacity for some slight inheritable variability" (*The Origin of Species* (1958), 449; Stauffer 1987, 224) and his claim that the universe cannot be conceived to be the result of chance, "that is, without design or purpose" (see Gillespie 1979, 140-45).

beyond it by insisting on a sensationist doctrine of perception, an atheistic view of the universe, and a materialistic view of finite existence.

7. *Nominalism*: This doctrine entails the rejection of "Platonic realism", according to which ideal forms or archetypes are inherent in the nature of things and play an explanatory role in the world. Nominalism – from the Latin *nomen,* meaning name – is the doctrine that the names for these forms are *merely* names, not pointing to entities that really exist in any sense. Neo-Darwinism is fully nominalistic, rejecting the realism about forms upon which the typological approach of Georges Cuvier (1769-1832) and other traditionalists was based. Ernst Mayr, for example, has said:

> "I agree with those who claim that the essentialist philosophies of Plato and Aristotle are incompatible with evolutionary thinking. [...] For the typologist, the type (eidos) is real and the variation an illusion, while for the populationists (evolutionists) the type (average) is an abstraction and only the variation is real". (1970)

The nominalism of Neo-Darwinism is entailed by its atheism, because only a universal agent could give ideal forms a home and render them causally effective. (I allude here to Whitehead's "ontological principle", which stipulates that nonactual things cannot exert agency and cannot even exist apart from being embodied in actual things.)

We turn now to some derivative scientific doctrines. These differ from the basic scientific doctrines by being derivative from the metaphysical doctrines, that is, from naturalismsam.

1.3 Four derivative scientific doctrines

8. *Gradualism*: This doctrine stipulates that *macroevolution proceeds gradually, through a step-by-step process comprised of very tiny steps.* As Darwin famously said:

> "Natural selection acts only by the preservation and accumulation of small inherited modifications, each profitable to the preserved being. (...) [N]atural selection

> [will] banish the belief of the continued creation of new organic beings, or of any great and sudden modification of their structure". (1958, 100)

Richard Dawkins has recently reaffirmed this approach. Asking how living things, which are "too improbable and too beautifully 'designed' to have come into existence by chance", did come into existence, Dawkins says:

> "The answer, Darwin's answer, is by gradual, step-by-step transformations from simple beginnings. (…) Each successive change in the gradual evolutionary process was simple enough, *relative to its predecessor,* to have come into existence by chance" (1987, 43).

This doctrine went against the traditional typological view, according to which tiny changes would result in incoherent, unviable organisms. According to Cuvier, who articulated this typological position, the principle of the correlation and interdependence of parts rendered the evolution of one species into another improbable. As John Brooke summarizes Cuvier's position: "There simply could not be a gradual accumulation of variation in any one part, unless all could change in concert. And that, for Cuvier, was simply too fanciful" (1991, 246).

Although gradualism is a scientific doctrine in the sense of being empirically testable, it is not scientific in the sense of being based on empirical evidence. The fossil record, which reveals almost nothing but well-defined types, with few intermediate varieties, led Thomas Huxley, in response to Darwin's acceptance of the dictum "nature does not make jumps"[3] to say: "You have loaded yourself with an unnecessary difficulty in adopting *natura non facit saltum* so unreservedly" (1901, 176).

Huxley, however, failed to understand that the difficulty was not "unnecessary", given Darwin's metaphysical principles. As Robert Wesson points out, "Darwin insisted on gradualism as the essence of naturalism and the repudiation of divine intervention" (1991, 38).[4] For Darwin, as Howard Gruber put it, "nature makes no jumps, therefore if something is found in the world that appears suddenly, its origins must be supernatural" (1981, 125-26).

---

[3] *The Origin of Species* (1958) 181, 191, 256, 435.

[4] As Gillespie notes, Darwin considered any suggestion of evolution *per saltum* (by jumps) to be a disguised appeal to miraculous creation (1979, 82).

Richard Dawkins, endorsing Darwin's position, says:

> "In Darwin's view, the whole *point* of the theory of evolution by natural selection was that it provided a *non*-miraculous account of the existence of complex adaptations. For Darwin, any evolution that had to be helped over the jumps by God was not evolution at all. (…) In the light of this, it is easy to see why Darwin constantly reiterated the *gradualness* of evolution" (1987, 249).

Darwin's rejection of saltations, however, did not reflect merely his rejection of miraculous, supernaturalist interruptions. It also reflected his denial of *any* form of (ongoing) theistic influence whatsoever.

The doctrine of nominalism is here the link. Anti-nominalists, affirming the real existence of forms or archetypes in the nature of things, could suppose that they might serve as "final causes" or "attractors", so that the jump from one coherent type to another would not be entirely accidental. They could, thereby, believe that it might occur occasionally. This influence could be intelligible, however, only if the forms or archetypes are thought to be rendered efficacious by a divine actuality.

9. *The Reduction of Macroevolution to Microevolution*: According to this doctrine, all macroevolution is to be understood entirely in terms of the processes involved in microevolution, which can be observed in the laboratory. Douglas Futuyma declares that "the known mechanisms of evolution [provide] both a sufficient and a necessary explanation for the diversity of life" (1979, 449). Richard Dawkins says that Darwinism "has no difficulty in explaining every tiny detail" (1987, 302). Walter Bock, a more circumspect Neo-Darwinist, admits that this has not been *shown*, saying: "One of the major failures of the [Neo-Darwinian] synthetic theory has been to provide a detailed and coherent explanation of macroevolution based on the known principles of microevolution" (1979, 20). Even Bock believes, however, that macroevolution can *in principle* be explained in terms of microevolution.

10. *Random Variations and Natural Selection as the Sole Principles of Evolution*: According to this doctrine, all the present species of life have come about through evolutionary descent from the first form(s) of life *solely* through natural selection operating upon random (or chance) variations.

The term “random” (or “chance”) does not imply that that evolution is not fully determined.[5] It means, instead, that mutations “random” in the sense that they are *not biased in favor of the adaptation of the organism to its environment*. In Niles Eldredge’s words, “mutations are random with respect to the needs of the organisms in which they occur” (1995, 133). As Stephen Jay Gould put it, genetic variation is “not preferentially directed towards advantageous features” (1981, 325). Randomness in this sense can hence be called “$\text{randomness}_{\text{na}}$” – with “na” standing for both “not advantageous” and “not adaptational”.

11. *Evolution as Wholly Undirected*: Variations are said to be random in an even stronger sense. In Gould’s words, variation is random in the sense of being wholly “undirected”, of “aris[ing] in all directions” (Gould 1982, 79; 1983, 138). Randomness in this sense, which we can call “$\text{randomness}_{\text{eps}}$” (for “every possible sense”), says that besides not being directed towards adaptation to the immediate environment, variations are also random in every other possible sense of the term. They are not biased toward, for example, the production of beauty, or greater complexity, or richer experience. This stronger meaning of randomness is intended to rule out the idea of *any type* of cosmic directivity. This idea, that evolution is completely “undirected”, was declared by Gould to be “the central Darwinian notion” (Gould 1982, 38).

1.4. Two religious-moral implications

12. *The Universe as Meaningless*: One of the things modern evolutionary biology teaches us, says Neo-Darwinist William Provine, is that “[t]he universe cares nothing for us. (…) Humans are as nothing even in the evolutionary process on earth. (…) There is no ultimate meaning for humans” (1988, 64-66, 70). Gould agreed, saying that we have to create our own meaning, because there is none in nature (1977, 13; 1982, 83; 1983, 93).

---

[5] These terms misled, for example, Holmes Rolston, who wrote that mutations are random in the sense of being “without necessary and sufficient causal conditions”, thereby completely contingent (1987, 104, 91).

13. *The Universe as Amoral*: This is the doctrine that universe contains no moral norms. Provine says that modern evolutionary biology "directly implies that there are no inherent moral or ethical laws" (1988, 64-66). Gould agreed, saying that "there is no 'natural law' waiting to be discovered 'out there'" (1992, 118).[6]

Insofar as Neo-Darwinists believe that to accept evolution is to accept religious-moral implications, they are claiming that we cannot have evolution without tears, that is, without accepting the conclusion that human lives, including all our finest achievements – such as the discovery of the evolutionary nature of our universe – are ultimately meaningless. We turn now to Intelligent Design, which finds this assessment of human life unacceptable.

2. Intelligent design's criticism of Neo-Darwinism

Advocates of Intelligent Design (ID), equating Neo-Darwinism with this maximal version of it, typically attack it on three grounds: It embodies a false metaphysics; this false metaphysics leads to scientific inadequacy; and this false metaphysics is morally and religiously deleterious. In discussing these criticisms, I will rely primarily on the writings of William A. Dembski, the author of *Intelligent Design* (1999) and *The Design Revolution* (2004).

2.1. Neo-Darwinism's false metaphysics

Dembski, using "naturalism" as the inclusive name for the metaphysical doctrine of Neo-Darwinism, says that the overall goal of ID is "overturning naturalism" (*Intelligent Design* (henceforth *ID*), 14). As to exactly what Dembski means by "naturalism", some definitions – such as the doctrine that "nature is all there is" (ibid. 100) – focus on Neo-Darwinism's athe-

[6] Helena Cronin reflects a similar position in saying that, "Man's inhumanity to man may indeed make countless thousands mourn. But it is man's *humanity* that gives Darwinians pause. (…) Human morality (…) presents an obvious challenge to Darwinian theory" (1991, 325).

ism. Some definitions emphasize the idea of "the world as a closed nexus of natural causes" in which "God plays no role" (ibid. 46, 99). This idea is alternatively stated as the idea that everything can be explained in terms of "undirected natural causes", so that "intelligent causes" are unnecessary (ibid. 46, 120). Still others emphasize that the world is "fundamentally an interacting system of mindless entities" so that mind is "an emergent property of suitably arranged mindless entities" (ibid. 232).

By putting Dembski's various definitions of naturalism together, we can see that ID is directed against naturalism$_{sam}$, especially its atheism and materialism. (Indeed, Dembski even says that what I call naturalism$_{sam}$ is essentially same as the naturalism he rejects (*The Design Revolution* (henceforth *DR*), 172). ID's critique of Neo-Darwinism is scientifically interesting, however, only when advocates of ID argue that Neo-Darwinism, because of its metaphysical naturalism, is scientifically inadequate.

2.2. Neo-Darwinism's scientific inadequacy

Dembski claims that the starting point of ID's critique is not Neo-Darwinism's incompatibility with Christian faith but its "failure as an empirically adequate scientific theory" (ibid. 112). Although that is a dubious claim – the starting point seems instead to be Neo-Darwinism's view of the universe as amoral and meaningless – ID advocates do level serious charges against the scientific adequacy of Neo-Darwinism. I will mention four.

One problem with Neo-Darwinism, says Dembski, is that by virtue of regarding mindless entities as fundamental, it must regard intelligence as "an evolutionary byproduct", but that this attempt "to reduce intelligence to natural mechanisms cannot succeed" (ibid. 22).

A second problem is that although Neo-Darwinism claims that all novel developments can be explained in terms of natural selection acting on random variations, this remains a mere claim. There are, Dembski quotes University of Chicago molecular biologist James Shapiro as saying, "no detailed Darwinian accounts for the evolution of any fundamental biochemical or cellular system, only a variety of wishful speculations" (ibid. 214, Shapiro 1996).

Third, there are good reasons to believe that this lack is not accidental – that organisms embody "irreducible complexity" and that Darwinian pathways to such complexity are impossible in principle (*DR* 293-98, *ID* 147-49).

Fourth, the fossil record, which suggests that new species appear suddenly and then remain basically the same for a long period, conflicts with the gradualism required by Neo-Darwinism.[7]

2.3. Neo-Darwinism's religious-moral destructiveness

ID's religious-moral critique focuses primarily on the atheism of Neo-Darwinism and its implication that the evolutionary process is entirely undirected. According to this view, as Dembski puts it, human beings "are not the crown of creation, (…) not creatures made in the image of a benevolent God" but "an accident of natural history" (*DR* 22). To illustrate this view, Dembski quotes George Gaylord Simpson's statement that "man is the result of a purposeless and natural process that did not have him in mind" (*ID* 117, Simpson 1967, 345). ID's critique, more generally, is aimed at Neo-Darwinism's nihilistic conclusion that the universe is meaningless and devoid of moral norms.

Robert Pennock, one of the chief critics of ID, believes, as I do, that this belief that Neo-Darwinism leads to meaninglessness provides the primary motivation behind the ID movement (1999, 311-27). Pennock suggests, therefore, that defenders of Neo-Darwinian evolution can "assuage the crisis of values that drives the creationist crusade" by "reassuring […] audiences that evolution does not imply atheism or moral nihilism" (ibid. 338). We can, in other words, have Neo-Darwinian evolution without tears.

Pennock's reassurance, however, is problematic. For one thing, although Pennock, equating Neo-Darwinism with its strictly scientific doctrines, says that "[n]owhere in evolutionary theory does it say that God does not exist", he also says that "God is not necessary to explain the modification of species one into another" and hence "to explain the biological

[7] This point has been developed most fully in Michael Denton's *Evolution: A Theory in Crisis*.

world" (ibid. 333). Pennock thereby supports ID's point that the only "God" allowed by Neo-Darwinism is one that is superfluous (*ID* 111-12).[8] Also, although Pennock may equate Neo-Darwinism with its scientific doctrines, some of its most visible advocates, as we have seen, emphasize its metaphysical doctrines, including its atheism, and explicitly draw nihilistic conclusions – conclusions that are supported by many leading moral philosophers, who say that atheism does imply a meaningless universe devoid of objective values (Mackie 1977, Harman 1977, Williams 1985, Rorty 1989).[9]

3. Could ID serve as a bridge between the religious and scientific communities?

These criticisms of Neo-Darwinism's metaphysical worldview, along with the scientific and religious-moral doctrines derivative there, imply changes in evolutionary theory that would make it compatible with Christianity and other religions. Accordingly, if ID is also acceptable from the perspective of the scientific community, it could indeed be, as Dembski proposes, a "bridge between science and theology".[10]

Dembski claims that there is no good reason for the scientific community to consider ID unacceptable. He seeks to make this claim plausible primarily by distinguishing ID from the position known as "scientific creationism" (or "creation science"). The most obvious difference, Dembski says, is that ID "has no prior religious commitments". In particular, besides not regarding the Genesis account as scientifically accurate, ID does *not*, Dembski claims, identify the intelligent cause responsible for nature's design as "a supernatural agent" who "create[d] the world out of nothing" (*ID* 247, *DR* 41). The charge that ID is committed to supernaturalism, he adds, is a "red herring" (*DR* 190). Although ID's designer "is compatible with the creator-God of the world's major monotheistic religions", Dembski

[8] See also Griffin 1998, or my critique of Phillip Johnson and Alvin Plantinga in Griffin 2000, 45-55.

[9] For commentary, see Griffin 2004a, 2005.

[10] This is the subtitle of Dembski's *ID*, except that Dembski presents it, less modestly, as *the* bridge.

says, it is "also compatible with the watchmaker-God of the deists [and] the Demiurge of Plato's *Timaeus*" (ibid. 44). Spelling out the implications of ID's lack of commitment to supernaturalism, he says that it "is not an interventionist theory" and does not presuppose miracles (ibid. 179, *ID* 107). Therefore, Dembski suggests, ID is a "third way" between supernaturalistic creationism and Neo-Darwinian naturalism, which could in principle be acceptable to both the scientific and the religious communities (*DR* 26-27).

Neo-Darwinist critics of ID, by contrast, portray it as anti-scientific.[11] Some of this criticism depends, to be sure, on regarding Neo-Darwinism, with its naturalism$_{sam}$, as *the* scientific view, so that to reject Neo-Darwinism is *ipso facto* to be anti-science. But part of this criticism reflects the fact that ID rejects not merely naturalism$_{sam}$ but also naturalism$_{ns}$.

For example, in spite of Dembski's insistence in places that ID does not necessarily require supernaturalism, he repeatedly indicates otherwise. Although he says that process theism, with its naturalism$_{ns}$ – which he calls "antisupernaturalist naturalism" – is compatible with ID, he declares it to be "incompatible with Christian theism", primarily because "Christian theology [...] regards the doctrine of *creatio ex nihilo* [...] as nonnegotiable" (*DR* 173, 174). Dembski goes on for several pages in this vein without explaining why this discussion is relevant if ID has no prior religious commitments. He even concludes by taking back his acknowledgment that process theology is compatible with ID, suggesting that only a God who creates *ex nihilo* can be a designer in the true sense of the term (ibid. 176).

Likewise, in spite of Dembski's claim that ID "does not require miracles", he says that "in a naturalized world that positively excludes miracles, design becomes increasingly implausible" (*ID* 51). He then proceeds to defend the reality of miracles, which occur when "God over-rules the inherent capacities of an entity, endowing the entity with new capacities" (ibid. 67) as in "the bodily resurrection of Jesus Christ" and "the virgin birth" (ibid. 42-43).[12]

In addition to rejecting naturalism$_{ns}$, Dembski appears to question the doctrine of macroevolution, or "common descent", which Darwin consid

[11] Pennock, in fact, has a chapter entitled "Burning Science at the Stake".

[12] See also *ID* 66.

ered the primary issue and which, as Pennock points out, evolutionists usually have in mind when they speak of "the fact of evolution" (Pennock 1999, 56, 57). Dembski, while pointing out that some ID advocates accept macroevolution, as least provisionally, insists that the discussion should remain open (*ID* 250). That he in fact accepts the special creation of each species is suggested, moreover, by some of his statements, such as his comment that the Christian God, being omnipotent, is capable of "creating species from scratch" and his statement that he regards CSI (complex specified information) as emerging "through discrete insertions over time" (*DR* 173, *ID* 171). It would seem, therefore, that Dembski favors the doctrine sometimes called "progressive creationism", which accepts the standard evolutionary timetable but sees each new species as separately created.

Accordingly, given ID's supernaturalism, which allows it to reject macroevolution in favor of progressive creationism, it does not provide a third way beyond Neo-Darwinism and creationism that could in principle be accepted by the scientific community. Such a third way, however, can be provided on the basis of Whitehead's philosophy.

4. Whiteheadian evolution as a third way

I will explain how Whiteheadian evolution provides a third way by employing the 13 doctrines of Neo-Darwinism outlined earlier.

Doctrines 1-4

Whiteheadian evolution accepts microevolution, macroevolution, naturalism$_{\mathrm{ns}}$, and ontological uniformitarianism. There can be, accordingly, no credible claim that Whiteheadian evolution, because of its rejection of the other doctrines of Neo-Darwinism, is somehow in principle unacceptable from a scientific standpoint.

Doctrines 5-7

Neo-Darwinism's naturalism$_{\mathrm{sam}}$, which is embodied in its doctrines of positivism-materialism, atheism, and nominalism, are replaced in Whiteheadian evolution by naturalism$_{\mathrm{ppp}}$, which symbolizes a *prehensive* doctrine of perception, a *panentheistic* doctrine of the universe, and a *panexperientialist* doctrine of actuality.

Panexperientialism solves one of the biggest problems of evolutionary theory – how the human mind, or animal minds more generally, could have evolved out of "matter" without supernatural assistance.

Panentheism combines the major concerns of Neo-Darwinists and ID advocates. On the one hand, as a form of naturalistic theism, panentheism provides an ontology in which any divine interruption of the universe's fundamental causal relationships is impossible in principle, because these causal relationships belong to the metaphysical fabric of the universe, and God is not "an exception to all metaphysical principles" but "their chief exemplification" (Whitehead 1979, 343).

On the other hand, this panentheistic doctrine, unlike the kind of "naturalistic theism" rejected by ID theorists, does not imply the deistic or even semi-deistic view once God had created the physical universe, God no longer exerted any variable causation. Rather, there is divine influence in every event and the content of this influence can vary enormously. (The divine aim for particular creatures is always for the best possibility open to them, and the possibilities that humans can actualize are radically different from the highest possibilities open to electrons or squirrels, and even the best possibility open to different human beings in different contexts will differ greatly.) This idea does not, however, contradict naturalism$_{ns}$, according to which the world's fundamental causal principles are never interrupted, because this divine influence is an essential part of these fundamental causal principles.

The prehensive doctrine of perception, according to which sensory perception is derivative from a more fundamental, nonsensory mode of perception, explains how human beings (along with other organisms) can directly experience divine influence.

Doctrines 8-11

Given the idea that every event is subject to divine influence – an influence that Whitehead calls an "initial aim" – the Neo-Darwinian doctrine that evolution is entirely undirected is rejected, along with the claim that macroevolution is fully understandable in random variations and natural selection.

From a Whiteheadian perspective, moreover, there is no *a priori* reason to assume that evolution would always proceed in a gradualistic manner. Relevant here is the Whiteheadian rejection of nominalism in favor of the idea that eternal forms ("eternal objects") exist in the "primordial nature" of the divine actuality, which presents them, by means of initial aims, as possibilities to be actualized. Accordingly, instead of having to choose between the standard options – either supernaturally produced saltations or steps tiny enough to have come about entirely by chance – we have a third option: creatures responding to lures to move from one coherent type to another.[13]

Doctrines 12-13
Whiteheadian evolution is nihilistic neither morally nor religiously. The eternal forms contained in the primordial nature of God include normative possibilities, which we prehend as moral norms (Griffin 2004b). Also God, in addition to this primordial nature, has a "consequent nature", meaning that everything that occurs in the world is everlasting by virtue of being taken into the divine experience. Whiteheadian evolution even allows for the possibility of life after death.[14] On the basis of Whitehead's philosophy, accordingly, we can indeed have evolution without tears.

Even with this very brief sketch of Whiteheadian evolution, accordingly, one can see that it does indeed present a third way – one that, unlike Neo-Darwinism, is acceptable from a religious-moral perspective and one that, unlike Intelligent Design, is acceptable from a scientific perspective.

5. Teaching evolution in public schools

A final question is what the perspective provided above would suggest about the embattled question of teaching evolution in public (tax-supported) schools. Most Neo-Darwinists evidently believe that Neo-Darwinism, at least as defined by doctrines 1-4 and 8-11, should be taught. Advo-

---

[13] See Griffin 2000, Chap. 8 "Creation and Evolution", esp. 302-08.
[14] See Griffin 2001, Chap. 6 "Evil, Evolution, and Eschatology"; 2000, 237-40; 2008.

cates of ID complain, however, that this involves teaching metaphysics under the guise of science, since doctrines 8-11 presuppose naturalism$_{sam}$, so that atheistic materialism is being implicitly taught as vouchsafed by science. These ID advocates argue that ID should be taught instead of, or at least in addition to, Neo-Darwinian evolution.

The above analysis suggests an approach that differs from both of these recommendations.

Insofar as evolution is taught as scientifically established fact, it would be limited to doctrines 1 and 2.

With regard to doctrines 3 and 4, students would be taught that they are accepted almost universally as necessary presuppositions of science, but that some philosophers, scientists, and theologians (such as Whiteheadian process theologians) accept naturalism$_{ns}$ as reflecting the nature of reality, whereas others regard it as merely a methodological convention of the scientific community.

With regard to the Neo-Darwinian interpretation of evolution as embodied in doctrines 5-11, students would be taught that although it has long been the dominant view in the scientific community, it is not universally accepted, even by those who accept doctrines 1-4; that metaphysical doctrines 5-7, which imply naturalism$_{sam}$, are neither proved nor required by science; that the derivative scientific doctrines 8-11, while accepted by Neo-Darwinists, have much less empirical support than do microevolution an macroevolution as such and are grounded more in naturalism$_{sam}$ than in empirical evidence.

With regard to the religious-moral doctrines 12-13, students would be taught that they, being rooted solely in doctrines 5-11, have no scientific support whatsoever, so are in no way implied by the acceptance of an evolutionary view of the universe.

Finally, students would be taught that the Neo-Darwinian claim that evolution is entirely undirected is challenged by theistic interpretations, some of which affirm naturalism$_{ns}$.

REFERENCES

Bock, W. (1979). "The Synthetic Explanation of Macroevolutionary Change: A Reductionistic Approach". In: *Bulletin of the Carnegie Museum of Natural History*, 18, pp. 20-69.

Brooke, J. (1991). *Science and Religion: Some Historical Perspectives*. Cambridge: Cambridge University Press.

Cronin, H. (1991). *The Ant and the Peacock*. Cambridge: Cambridge University Press.

Darwin, C. (1872/1958). *The Origin of Species*. New York: Mentor Books.

Dawkins, R. (1987). *The Blind Watchmaker: Why the Evidence of Evolution Reveals a Universe without Design*. New York and London: Norton.

Dembski, W. (2004). *The Design Revolution: Answering the Toughest Questions about Intelligent Design*. Downers Grove, Ill.: InterVarsity Press.

—— (1999) *Intelligent Design: The Bridge between Science and Theology*. Downers Grove, Ill.: InterVarsity Press.

Denton, M. (1991). *Evolution: A Theory in Crisis*. London: Burnett Books.

Eldredge, N. (1995). *Reinventing Darwin*. New York: Wiley.

Futuyma, D. (1979). *Evolutionary Biology*. Sunderland, Mass.: Sinauer.

Gillespie, N. (1979). *Charles Darwin and the Problem of Creation*. Chicago: University of Chicago Press.

Gould, S. (1992). "Impeaching a Self-Appointed Judge". In: *Scientific American* (July), pp. 118-21.

—— (1983). *Hen's Teeth and Horse's Toes*. New York: W. W. Norton.

—— (1982). *The Panda's Thumb*. New York: W. W. Norton

—— (1981). *The Mismeasure of Man*. New York: W. W. Norton.

—— (1977). *Ever Since Darwin*. New York: W. W. Norton.

Griffin, D. (2008). "Process Eschatology". In: Walls, J. L. (ed.). *The Handbook of Eschatology*. Oxford: Oxford University Press, pp. 296-310

——(2005). "Theism and the Crisis in Moral Theory: Rethinking Modern Autonomy". In: Allan, G. and Allshouse, M. F. (eds.). *Nature, Truth, and Value: Exploring the Thought of Frederick Ferré*. Lanham, Md: Lexington Books, pp. 199-220.

——(2004a). "Morality and Scientific Naturalism: Overcoming the Conflicts". In: Hackett, J. and Wallulis, J. (eds.). *Philosophy of Religion in the New Century: Essays in Honor of Eugene Thomas Long*. Dodrecht: Kluwer Publications, pp. 81-104.

——(2004b). "Feeling and Morality in Whitehead's System". In: Helmer, C. with Suchocki, M.; Quiring, J. and Goetz, K. (eds.). *Schleiermacher and Whitehead: Open Systems in Dialogue*. Berlin and New York: Walter de Gruyter, pp. 265-94.

——(2001). *Reenchantment without Supernaturalism: A Process Philosophy of Religion*. Ithaca: Cornell University Press.

——(2000). *Religion and Scientific Naturalism: Overcoming the Conflicts*. Albany: State University of New York Press.

——(1998). "Christian Faith and Scientific Naturalism: An Appreciative Critique of Phillip Johnson's Proposal". In: *Christian Scholar's Review*, 28/2 (Winter), pp. 308-28.

Gruber, H. (1981). *Darwin on Man: A Psychological Study of Scientific Creativity* (2nd edition). Chicago: University of Chicago Press.

Harman, G. (1977). *The Nature of Morality: An Introduction to Ethics*. New York: Oxford University Press.

Huxley, L. (ed.) (1901). *Life and Letters of Thomas Henry Huxley*, 2 vols. London: Macmillan; New York: A. Appleton, Vol. II.

Mackie, J. (1977). *Ethics: Inventing Right and Wrong*. New York: Pelican Books.

Mayr, E. (1970). *Populations, Species and Evolution*. Cambridge: Harvard University Press.

Pennock, R. (1999). *Tower of Babel: The Evidence against the New Creationism*. Cambridge, Mass.: MIT Press.

Provine, W. (1988). "Progress in Evolution and Meaning in Life". In: Nitecki, M. (ed.) *Evolutionary Progress*. Chicago: University of Chicago Press, pp. 49-74.

Rolston, H. (1987). *Science and Religion: A Critical Survey*. Philadelphia: Temple University Press.

Rorty, R. (1989). *Contingency, Irony, and Solidarity*. Cambridge: Cambridge University Press.

Shapiro, J. (1996). "In the Details ... What?" (review of Michael Behe's *Darwin's Black Box*). In: *National Review*, September 16, pp. 62-65.

Simpson, G. (1967). *The Meaning of Evolution*. New Haven: Yale University Press.

Stauffer, R. (ed.) (1987). *Charles Darwin's Natural Selection: Being the Second Part of His Big Species Book Written from 1856 to 1859*. Cambridge: Cambridge University Press.

Wesson, R. (1991). *Beyond Natural Selection*. Cambridge, Mass.: MIT Press.

Whitehead, A. (1967). *Science and the modern world*. New York: Free Press.

Williams, B. (1985). *Ethics and the Limits of Philosophy*. Cambridge: Cambridge University Press.

Covariance and Evolution

ROBERT J. VALENZA

This paper considers, within an abstract framework, how the objects of the world become known to us, not by sensation, but by a cognitive process that implicitly invokes two general notions, one from mathematics, the other from physics, that have only been clearly understood relatively recently. Our plan is to give brief introductions to each and to their connections with ontology, and then to explore a hypothetical connection with evolution and some of the problems of philosophy of mind. This latter business is, to say the least, liberally speculative, but here we may claim to be working in the spirit of Nagel (1986; see especially the early chapters) in assuming that something well out of the box has been wanting in our approach to metaphysics. Indeed, this well-out-of-the-box thinking might find its foundations in the process metaphysics of Alfred North Whitehead.

Whiteheadian metaphysics in particular and dual aspect theories in general allow that reality coheres in nodes that admit experience. Such entities may exhibit two modes: all carry a subject-centered phenomenal aspect, and some of the more complex ones also manifest a perspective-free epistemological mode. Each mode supplies part of the basis for a rational ontology, but it is the latter that affords the possibility of a fully sharable worldview, and, in particular, a community-wide ontological deployment.

We claim that at the heart of the technical mechanism by which subjects distinguish world objects is the *principle of covariance*, a generalized conception of symmetry that emerged early in twentieth-century physics. Covariance, we further argue, operates in connection with the so-called *quotient construction* from elementary mathematics – an ontological move with which most of us are in fact implicitly familiar from surprisingly early in our formal education. In light of these claims, we explore the hypothesis, suggested by the history of the world and science, that physical and phenomenological covariance play a key role in the development of life forms of greater and greater complexity. Explanations of this dynamic might include, among others, Whitehead's theory of the relationship of God to actual entities in general. As a consequence of this hypothesis, physical reductionism phrased in terms of a third-person ontology must be incomplete either at the level of ontogeny or phylogeny insofar as the very notion of covariance is necessarily grounded in symmetries founded upon first-person references.

1. The quotient construction

Let us begin with something well understood in mathematics that apparently has not received much attention in metaphysics. We shall proceed by first introducing a familiar example that is key and then moving on to a more abstract discussion.

Most of us would have been well prepared by the age of ten or so to affirm numerical identities of the following sort:

$$\frac{1}{2}=\frac{6}{12}$$

Let us pose two questions in connection with this particular example: First, what exactly is being asserted by this equality? Second, how do we in fact affirm it?

The second question is easy and could have been disposed of by our ten-year-old former selves. Rising from a small desk, any one of us might have declared, "Cancel the sixes out of the numerator and the denominator of the right hand fraction and what remains is the left hand fraction". It is much less clear what we might have said in response to the first question at that age, but speaking from a more mature perspective, one might, still with a certain mathematical naiveté, want to say something like this: "There is this thing – a concept? something that actually exists? – called one-half. It has many representations, two of which are given in the displayed equation above. The equality merely states that these particular representations denote the same thing". Bravo! Despite the uncomfortable squirming to be seen among the future philosophers in our imaginary classroom, this is an astute answer – except for the ontological problem that underlies it.

The mathematical answer to the ontological problem – at least a relative one – is not all that complex, but it is deep insofar as it has an amazing universality.[1] We walk through this answer quite informally in the next few paragraphs. Familiarity with the attendant concepts, not technicalities, is all we shall need subsequently. The key is to identify a special kind of relation that may hold between two objects.

In general, we may think of a (binary) *relation* on a set S as a two-place predicate that accepts elements of S for its variables. Hence for ordinary integers a and b, here are some exemplary relations:

(i) $a = b$
(ii) $a \neq b$
(iii) $a < b$
(iv) $a \leq b$
(v) $a \mid b$

Perhaps only the last of these is unfamiliar; it denotes divisibility, as in four divides eight. In general a relation $\sim$ on a set S is called an *equivalence relation* if it satisfies the following three properties:

Reflexivity. For all a in S, $a \sim a$.
Symmetry. For all a and b in S, if $a \sim b$ then also $b \sim a$.
Transitivity. For all a, b and c in S, if $a \sim b$ and $b \sim c$, then also $a \sim c$.

Of the examples above, only equality qualifies as an equivalence relation, and indeed one can think of the whole notion of an equivalence relation as a generalization of equality.[2] Other familiar examples from elementary geometry include congruence and similarity.

We come now to what is, for our application, the most important feature of an equivalence relation. We state it first and then give a graphical illustration.

---

[1] Two points: With regard to complexity, this matter is routinely handled by foundational courses with names like Set Theory and Logic taught to second- or third-year undergraduates. My point in speaking of it as a relative answer to the ontological problem is that, as we shall see, the solution is only a solution given a prior class of bona fide objects.

[2] More explicitly, reflexivity fails for examples (ii), (iii); symmetry fails for (iii), (iv) and (v); transitivity fails only for (ii).

Let ~ be an equivalence relation on a set S. Then ~ partitions S into a union of disjoint (that is, non-overlapping) subsets, each of which consists of all of the elements of S that are mutually equivalent with respect to the given relation.[3]

Let's work through this in the case of the set *S* illustrated below, which consists of a collection of various shapes in various shading patterns. (Each separately pictured element is considered distinct.)

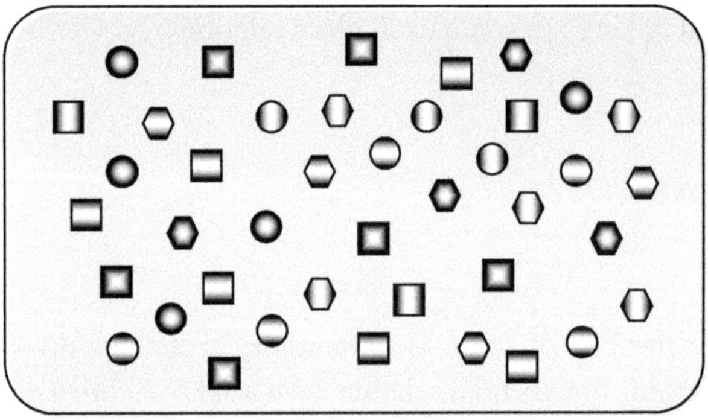

Fig. 1: A set *S* of shaded shapes

There are three distinct shapes, and three distinct shading patterns. Let us declare two elements of this set equivalent if they have the same shading pattern. This is clearly an equivalence relation in the sense defined above. We may then collect all of the similarly shaded objects into three classes as shown in Fig. 2. We see that the original set has been partitioned into three subsets that do not overlap. Each of these three subsets merely collects together every element in the original set of a given shading pattern, and thus every element of the original set lies in exactly one of these pieces of the

[3] This statement derives an associated partition from an equivalence relation. This association is trivially invertible: given any partition of a set *S* we can derive an equivalence relation ~ by declaring $a \sim b$ for two elements *a* and *b* in *S* whenever they lie in the same component of this partition. For example, given the partition illustrated in Fig. 2, we can reconstruct the equivalence relation that generated it by declaring two objects equivalent if they lie in the same swath. Note that this indeed reconstructs the original relation in the sense that the same pairs of objects satisfy it as before, but does not directly reconstruct its semantics.

partition. This is to say precisely that S is the disjoint union of these three subsets.

This last figure is critically important because it supports something that, speaking now in a philosophical vein, may be considered an ontological move. The original set S, which by the way contains 41 elements, each of which is a single shaded shape, has, via this equivalence relation, given rise to a new set that consists of only three elements: the three indicated components of the partition. These components are themselves subsets of S, and each is just the assembly of all the instances of a given shading pattern.

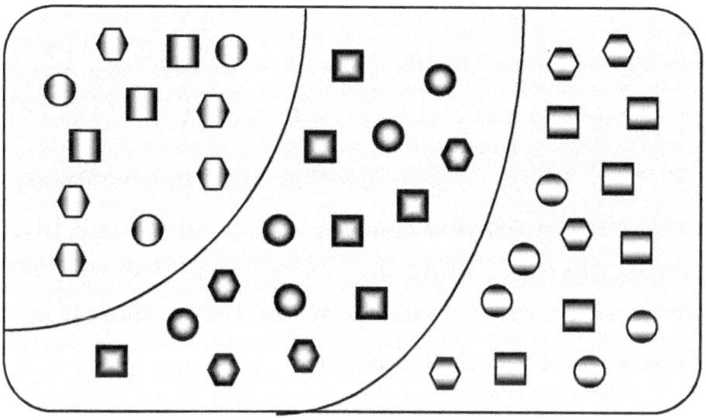

Fig. 2: A set partitioned by an equivalence relation

The general terminology for all of this is the following: The elements of the partition defined by an equivalence relation ~ on a set S are called, with evident good reason, the *equivalence classes* of S with respect to the relation ~. The set of all such equivalence classes is called the *quotient set* of S by the relation ~ and often denoted $S / \sim$. The terminology and notation are both astute because this quotient construction divides the original set into the pieces defined by the equivalence classes.

To summarize the graphical example above in the terminology just introduced, the quotient set in this case consists of the three equivalence classes demarcated in Fig 2, corresponding to the three shading patterns. *One cannot overemphasize that the original set and the corresponding quotient consist of things of distinct types.* The elements of the quotient are in fact certain subsets of the original set, and hence there is a definite aspect of ontological ascent to the quotient process.

Let us now revisit the matter of fractions to see how all of this applies. Consider the set S of all ordered pairs (a, b) of integers, where it is agreed that the second member of the pair shall never be allowed the value zero. Now consider the relation $\sim$ defined for these ordered pairs by the condition:

$$(a, b) \sim (c, d) \text{ if } ad = bc$$

It is easy to check that this is an equivalence relation on the set S. To get some handle on what this equivalence relation means, let us write out two distinct chains of equivalences, all trivial to check:

$$(1, 2) \sim (2, 4) \sim (3, 6) \sim (4, 8) \sim (5, 10) \sim (6, 12)$$
$$(2, 3) \sim (4, 6) \sim (6, 9) \sim (8, 12) \sim (10, 15) \sim (12, 18)$$

It would seem from these examples that this equivalence relation has something to do with ratios, *and that is exactly right!* But the point is that we are at the threshold of defining the notion of a ratio, or fraction, relative to the prior notion of an integer (or, more precisely, that of an ordered pair of integers). Here it is: define the fraction

$$\frac{a}{b} \quad (a \text{ and } b \text{ integers, } b \neq 0)$$

to be the equivalence class of the ordered pair (a, b) in S under the equivalence relation $\sim$ given above. Then a fraction, in this strict definition, is an equivalence class of ordered pairs of integers, and the statement that we examined above

$$\frac{1}{2} = \frac{6}{12}$$

is really the statement that 1/2 and 6/12 amount to the same equivalence class represented in different ways. This assertion in turn reduces to the evident truth of the arithmetic statement that $1 \times 12 = 2 \times 6$, which, we must note carefully, does *not* reference division or ratios in any way. It is well worth further noting here that the formal constructions of the integers from the natural numbers, the rational numbers from the integers, and the real numbers from the rational numbers are all formally achieved by equivalence relations and quotient sets.

We close this mathematical discussion of quotient sets with a question that may be described as either metaphysical, psychological or mathematical. We pose the question again with respect to our example of fractions, now defined via an equivalence relation on ordered pairs of integers: Is the notion of a fraction indeed prior or consequent to this construction? Historically, it is certainly prior, which is to say that the concept was there first. What, then, has been achieved with the formal construction and identification of the prior concept with this formalism? Obviously the answer is foundational and ultimately epistemological: we can now prove things about fractions as they have been constructed from integers (and ultimately from natural numbers) that we may have taken as merely axiomatic or intuitive before. But what, then, is a fraction really? Let me not pretend to answer this question except to say that everything we seem ever to have expected from fractions as fractions seems completely encoded in the formal construction, with its consequent epistemological safety.

2. Equivalence and ontology

This section is transitional. We shall look somewhat less abstractly at a particular kind of equivalence and its relation to a certain species of objects.

The four shapes in the figure directly below are called *quadominoes* and constitute a special case of a more general construction of figures called *polyominoes*, among which the common domino is the most familiar representative. A polyomino is a shape built from a number of squares joined to one another along complete edges. A quadomino is a polyomino built from exactly four squares. Our only point in introducing them here is to ask this: In what sense do the quadominoes in Fig. 3 exhaust all of the possibilities?

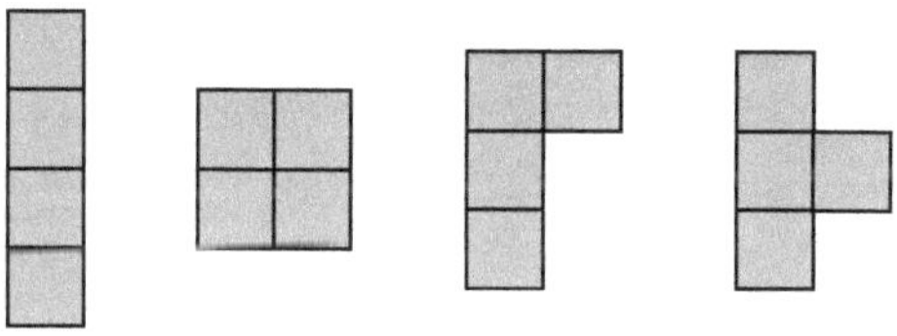

Fig. 3: The quadominoes

To sharpen this question, consider the two pairs of quadominoes in the following figure. In what sense do we consider the members of the left hand pair as representing the same quadomino? Well, certainly if one were to think of the two objects as rigid bodies in space, rotating either one by 180° would transform it into an exact copy of the other. Similarly, for the right hand pair, if we flipped (or reflected) either one horizontally, we would obtain the other.

Fig. 4: Two pairs of equivalent quadominoes

The point of this bit of analysis is only to indicate that there is a familiar kind of equivalence that we deal with all the time in assessing the sameness of shapes, and that is merely this: we say that two geometric objects have the same shape (or are congruent) if one can be transformed into the other by some series of *rigid motions*, which include rotations, reflections and translations.[4] In connection with the previous section, we want to stress that the relation "Figure *A* can be transformed into figure *B* by a series of rigid motions" is an equivalence relation. The distinct shapes then appear again as a quotient set, and our statement that Fig. 3 exhausts all possible quadominoes is really a statement about a quotient set.

Now, does this point extend more generally into metaphysics? I think the answer – construed less formally, of course – is that it does, and indeed at a primary level in questions of identity and substance (see Munitz 1971 and Wiggins 1980, for an introduction to these issues.) My point is perhaps an obvious one with regard to simple objects. For instance, just as equivalence under rigid motions leads me to assert that two quadominoes are indeed identical, a similar set of transformations through space and time helps with simple objects such as my copy of the book by Wiggins just cited. At my most cautious and conservative appraisal of the situation, I have no doubt that the volume I carried away from my bookshelf some ten minutes ago is the very same one that now rests on my desk to the right of

[4] To translate an object is to slide it through space so that all of its parts move in parallel.

my keyboard. Clearly my experiences of this object over this time span are hardly identical, but I amalgamate those experiences to posit a single object based on a putative set of transformations from experience to experience that mark certain elements as related in a manner that is characteristically symmetric, reflexive, and transitive. For instance, a certain brown and white visual pattern that bounds the spatial extent of this book, as I judge it from my particular perspective on the world, has moved in a continuous and evidently reversible path over the period of consideration. If two experiences of this pattern are marked as equivalent whenever they are linked by just such a path, I have an equivalence relation and a quotient object: my copy of the book. I have oversimplified here insofar as my actual experience of this object has not been merely visual, but if we imagine a silent video of ordinary events, we see the quotient operation I have described pretty much defining the objects on screen.

This paper is not the place to attempt a comprehensive theory of entification by equivalence, but in moving forward, I should like to point out that it is at least a plausible perspective for the following reasons:

1. The underlying mathematical model has been thoroughly vetted and clearly reflects useful insights into cognition and abstraction. In particular, the notion of an equivalence relation upon which the general construction of quotient sets rests is a direct generalization of equality, which is to say, identity.
2. The application to ordinary objects in ordinary circumstances is again analogous to something well understood in geometry, as indicated above.
3. Entification by equivalence is consistent with the discovery and deployment of the objects of science that are identified by conservation laws (e.g., mass, total energy and charge in classical physics; mass-energy in relativistic physics).[5]

---

[5] This is a bit technical and involves the so-called *universal property* of the quotient construction. We can paraphrase it informally as follows: Given a set S with an equivalence relation $\sim$ and a function defined on S that gives equal values on inputs that are equivalent with respect to $\sim$, the function is actually defined on the quotient set $S/\sim$. The idea that mass (or mass-energy) attaches to a persistent object (viewed as an equivalence class) then becomes a special case of this property.

4. Many of the technical arguments given in metaphysical theories of identity and substance, often arguments by continuity, are in fact applications of this general construction.
5. Quine's well known objection to the entification of properties (that individuation demands specificity of definition) might well be construed as a fundamental distinction between the components that coalesce by equivalence and the criteria by which this equivalence is defined (for instance, "No, that's my apple – this one's yours. Mine was green").[6]
6. The recognition of formal equivalence as a force in ontology seems, at least to me, enlightening vis-à-vis some of the now classic puzzles in identity theory (involving, for instance, split brains, rebuilt ships and forking roads) insofar as it tends to highlight the successes and failures of symmetry and transitivity in the criteria for identity.
7. Finally, there is a satisfying reciprocity between equivalence and covariance that we shall explore below.

One final note: The ideas we have considered here pertain to ontology as it distills out of experience. In that sense, it is the ontology of the individual, just as the *Tao Te Ching* in a similar sense represents a guide for the individual. When we pass out into the community, as we might from the Laotzu to Master Kong, something new immediately appears. In *The Analects*, perhaps this fundamental new concept is *ren*, perfect humaneness or reciprocity. In our setting, the key concept to be introduced is *covariance*, an aspect of our more intuitive notion of symmetry.

3. Symmetry and ontology

The idea of equivalence as it relates to ontology so far governs the subjective or interior perspective of an individual in the world. When we pass to the objective or exterior view of the world, to all that is perspective free, as Thomas Nagel might further characterize it, another principle operates: a kind of reciprocity between objects and perspectives, or ontology and epistemology. The main idea implicated is a generalization of our intuitive ge-

[6] I take no position on Quine's objection here, but see Henry and Valenza, 2001.

ometric idea of symmetry. Again, we ease into our main point by restricting ourselves at first to a sharply defined mathematical domain.[7]

Let us consider symmetry as it pertains to triangles, regular polygons and the circle. In what sense is it intuitively appealing to assert that an isosceles triangle is more symmetric than a scalene triangle (i.e., one that is not isosceles)? Or that an equilateral triangle is more symmetric than a merely isosceles triangle? And so on for squares, pentagons, hexagons, and ultimately for circles? The answer involves that same group of rigid transformations we met previously in connection with quadominoes, but at this point our interest is not in transforming one figure into another, but in leaving a given shape invariant. In this light, consider Fig. 5.

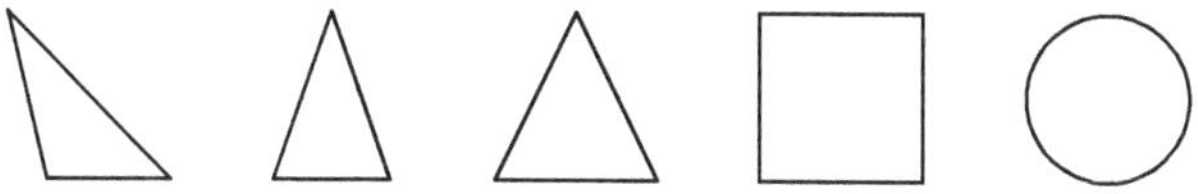

Fig. 5: Figures of increasing symmetry

We begin with the isosceles triangle, second from the left. Imagine that we flip it, like a pancake, horizontally and precisely through the perpendicular bisector of its base (not shown). The figure ends up right where it started, occupying exactly the same points in the plane. This particular reflection is suggestively called a *symmetry* of the isosceles triangle. (Note that this use of the word is, in this instance, technical, as opposed to the general notion of symmetry that we have used previously, and indicates a rigid motion that satisfies a particular property with respect to the figure.) One can see that no such symmetry exists for the scalene triangle, on the extreme left, which is to say that no reflection through any line whatsoever leaves this figure intact in the sense just indicated. Next consider the equilateral triangle, in the middle of the Fig. 5. It admits not one, but three reflections that constitute symmetries of this figure: reflection through the perpendicular bisector for any of the three sides will do. And there are more: the equilateral triangle admits an entirely new kind of symmetry arising from another kind of rigid motion, namely rotation. Rotations about its center by either 0° (i.e., doing nothing),

[7] Some of the points of this section and the next are made briefly in my Einstein article in Michel Weber's *Handbook of Whiteheadian Process Thought* (2008a).

120° or 240° move the figure back onto itself. Including the do-nothing rigid motion as a symmetry in all cases, we can now count the number of symmetries in these first three shapes: they are respectively one, two and six. Here we have the beginning of a definite measure of symmetry that respects our intuitive sense of the word. Continuing on, one can find eight symmetries for the square to the right of center, ten for the pentagon (not shown), etc. Finally, what about the circle? In this case a reflection through *any* line through the center and a rotation about the center by *any* angle both constitute symmetries; thus there are infinitely many symmetries for the circle, which gives satisfying technical support to our intuition that the circle is far more symmetric than any regular polygon.[8]

To make our point about the connection between symmetry and ontology, we now have to examine the same idea from a dual perspective, which turns out to be more abstract. In the preceding paragraph, we conceived of symmetry as being encoded in certain transformations that we might apply to a given figure. Implicit in this was the feature that the transformation operated from the same perspective; put more technically, the object was moved about against the background of a fixed coordinate system. In the view we put forth now, it is the coordinate system itself that will be transformed. We can understand this easily enough with the help of a new picture:

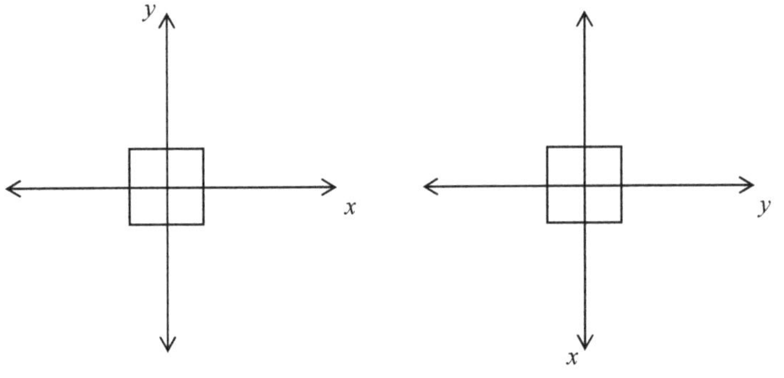

Fig. 6: Changing coordinates

[8] Single polygons and circles do not illustrate symmetries represented by our last class of rigid motions, the spatial translations. These, however, are nicely illustrated by tiling patterns stretched to infinity. For example, think of a checkerboard that goes on in all four directions forever. Shift this infinite board two squares to the right or left or up or down and the pattern lands on itself. A diagonal shift of the appropriate length will likewise serve.

In the figure above our square sits, on the left, in the center of a standard coordinate system for the Euclidean plane; the square is oriented so that its sides are parallel to the coordinate axes. We can imagine describing the square in this coordinate system by giving some sort of description of the coordinates that it occupies. For example, if a side of the square has length two, the points (1, 0), (0, 1), (-1, 0) and (0, -1) will be among these occupied points. Let us now shift our attention to the right hand part of the drawing. Here we should imagine the same square, undisturbed, but now viewed in a new coordinate system obtained from the old by rotating the axes clockwise by a right angle. If we were again to describe the square by giving a comprehensive listing of its coordinates in this new system, what would we find? The same set of coordinates as previously! Note that this assertion evidently fails if we rotate by only 45°, or by anything else that is not an integer multiple of a right angle. Thus we have a second perspective on the rotational symmetries of the square, and indeed on all of its symmetries: they are described by transformations of the coordinate systems that leave an analytic-geometric description of the square undisturbed. The same, of course, applies to the other figures we have considered above. The point is that seen in this way, these geometric symmetries are now reflected in coordinate transforms, or shifts in perspective, rather than in transformations applied to the objects themselves. Symmetry has become a kind of invariance under coordinate transforms, or, more colloquially, a kind of perspective invariance.

Our proposed connection between ontology and symmetry may now be made via this latter characterization of symmetry as perspective invariance. Given a collection T of coordinate transformations (for instance, those corresponding to the rigid motions we considered above), we may hold that the only legitimate objects of discourse are those that are invariant under the transformations of T. We shall return to the plausibility of this important statement shortly, but, for the moment just to illustrate it, consider the set T that consists of rotations by any integer multiple of a right angle. If we restrict our discourse to objects that remain invariant under T, we may speak of squares, octagons and circles, among other things, but not of triangles and ellipses, among *many* other things. Now if we expand T to include also rotations by angles that are multiples of 45 degrees, our corresponding ontology must surrender its squares, but will retain its octagons and circles. If we now expand further to *all* rotations, only the circles survive. The point is that as we demand more perspective invariance, we diminish our ontology.[9]

4. Covariance in mathematics and physics

We now pass from the notion of symmetry in the specific domain of geometry to a more general conception. Historically, it is fair to say that this generalization surfaced first explicitly in physics and later was captured more abstractly by mathematics; nonetheless, we shall consider the abstract mathematical formulation first.

The term *functoriality* refers to the preservation of abstract formalisms under transformation. To be more precise, imagine first a class C of objects, referred to in the current context as the *source class*, together with certain given (directed) relations among those objects, technically called *morphisms*. (For example, C might consist of the class of all two-dimensional manifolds – spheres, donuts and such – with the relations of interest being simply the continuous functions between such objects.) Next imagine a second class D of objects, the *target class*, with its own given set of special relations. (For example, D might consist of the class of abstract groups with the relations taken as group homomorphisms.) It is paramount to note that C and D may have little resemblance, so that C may arise in geometry and D in algebra. Finally, suppose that to every object *A* in C we

[9] By the way, taking a reciprocal approach from the other direction, as we did in our discussion of quotient sets, we might declare two objects identical if one can be obtained from the other by a transformation in our given collection *T*. (Here we again think of the transformation applying to the object, rather than to the coordinate system.) Thus in our example of polygons and circles, with respect to the collection *T* of *all* rotations, any two squares of the same size and centered at the origin may be taken as identical. Effectively this inflates the symmetries of the square to include all of *T*. Just as with the quadomino example, the size of the ontology and the number of symmetries vary inversely. Using this reciprocity, we might view the constructions of private and public ontological systems that we have considered – in the first case, transforming objects and in the second, transforming coordinates – as two aspects of essentially the same process.

can assign an object A' in D, and, moreover, to every relation R between two source class objects we can assign a relationship R' to the corresponding target objects; that is, every relation R between source objects, say, A and B is assigned to a relation R' between the corresponding target objects A' and B'. Such a complex and structured correspondence across classes is called a *functor* provided that certain other natural conditions are met. When a functor preserves the direction of source and target relations (a relation from A to B becomes a relation from A' to B'), it is called *covariant.* When it reverses the direction of the source and target relations (in this case a relation from A to B becomes a relation from B' to A'), we call it *contravariant.*

To the uninitiated, the preceding paragraph is probably intelligible taken one sentence at a time, but puzzling to the extent that one might doubt that such an abstract concept has any genuine relevance, even to mathematics.[10] This is not surprising when one considers that these ideas were only introduced in the mid twentieth century after millennia of accumulated and appropriately distilled mathematical experience (Eilenberg and MacLane 1945), but our only point here is to suggest just how basic and austere the notion of functoriality indeed is.

In the somewhat less austere, but still highly abstract context of physics, the preservation of form under transformation goes by the more particular name of *general covariance.* General covariance in physics is a cognate of mathematical covariance insofar as it maintains that the *form* of a physical law should likewise persist across admissible changes in coordinate systems. An early, implicit example of this might be Newton's laws of motion with respect to coordinate transforms that arise from one inertial frame of reference moving with constant velocity with respect to some other such frame. The rules that govern the coordinate and velocity changes between two such systems are called the *Galilean transformations.* One

---

[10] In this connection the examples from the previous paragraph are most relevant: A good deal of algebraic or geometric topology may, in the large, be considered as discovering functors from manifolds to groups, thereby converting continuous functions to group homomorphisms. The point is that often the corresponding group structures are far simpler than the original manifolds, but enough of the relations among manifolds are preserved by the functor that decisive information can be read off the associated group constructions.

may then observe that positions and velocities differ according to the frame of reference, but the *forms* of Newton's three laws remain the same. A more sophisticated and historically significant example arises in connection with Maxwell's equations, which govern classical electromagnetism. Einstein noticed that while these equations cannot retain their form under Galilean transformations, because, for instance, the speed of light is constant in all frames, they do retain their form under the *Lorentz transformations*. If we accordingly take the Lorentz transformations as correct, Newtonian mechanics must yield to the dynamics associated with the theory of special relativity. This led Einstein to one of the greatest ideas in the history of science: if one then elevates even further the principle that physical laws should exhibit covariance so to include not just shifts between inertial coordinate systems, but a more general class of transformations including mutually accelerated frames, one is led from special to general relativity and, in particular, to a whole new theory of gravity.

To view these scientific advances more abstractly, one might say that in consolidating the form of fundamental physical laws as they applied in an inertial frame, Einstein introduced a new set of symmetries in space-time in the same sense that one might plausibly maintain that the "discovery" of a circle would encourage the exploration of rotational symmetries in general. But just as symmetries exert an influence on ontological systems in the abstract example above, these new symmetries had implications for the objects of science, and certain classical quantities which were of use precisely because they were conserved in non-relativistic physics had to be redefined accordingly. For example, Newton's law of conservation of momentum is made invariant under Lorentz transformations by a redefinition of mass, which is no longer a constant attached to an object, but now dependent as well on its state of motion within a particular frame.[11] Mass is thus no longer conserved, but something else, mass-energy, is. Hence one of the fundamental notions that implicitly defined a physical object as a closed system appropriate for physical investigation and analysis had to yield to the requirements of a new kind of symmetry – again, we stress, viewed in the abstract as invariance under coordinate transformation – and the expansion of symmetries in turn has exerted itself on the objects of the

---

[11] See Resnick 1968, 111-17.

theory. Even more spectacularly, in the general theory of relativity the equivalence of gravitational and inertial mass, again an aspect of the requirement for general covariance, expands the symmetries of space under consideration vastly beyond the Lorentz transformations. Gravity then becomes an aspect of the structure of space itself as we pass from the "flat" world of classical or special relativistic physics into a universe with an intrinsic curvature. Here, too, ontology bows to symmetry.

In a most general sense, much of science, even beyond physics, can be seen as consonant with a covariance principle. Any atomic theory – even going back to Democritus – and perhaps reductionism in general, seeks to free our understanding of what happens here and now in this or that particular situation of all of the attendant particulars and to reduce it to time- and space-independent properties, which is to say, to features that are free of absolute coordinates or perspectives. To put this rather more vaguely, there is an implicit acknowledgement that the form of our scientific understanding of the world should be perspective free. But put this way, we can hardly be surprised at such a principle, because this is the very essence of objectivity.

5. Covariance and metaphysics

I suggested earlier in this paper that a cognitive being judges what should count as objects in its world by a process similar to a fundamental mathematical construction. This is not to endorse some form of idealism, but to suggest something of what we do, as individuals, to make coherent sense of the real world. Later I argued that, at least in science, the idea of general covariance, an extension of our intuitive idea of symmetry, influences the form and ontology of physical sciences in a grand, but still very formal sort of way. In this section we shall look at these matters from a more general metaphysical perspective.[12]

[12] I owe my first contact with the notion of covariance in the context of general metaphysics to Joachim Stolz and his splendid monograph *Whitehead und Einstein* (1995). Stolz argues that even though Whitehead never explicitly mentions covariance in his philosophical writings, it is the defining leitmotiv of the etiology of process thought (p. 156). Indeed, process metaphysics is characterized by a set of dynamical objects in

I am sitting at present in such a location that I can see both a yellow pad to the right of my keyboard and the direct rays of sunshine through a window. When I look into the sunlight and then look away, I see a dark spot, a negative after-image, where the light was concentrated. This spot has the kind of integrity, coherence and persistence such that I have no metaphysical hesitations about calling it a spot. A sequence of experiences has entified along the lines that I discussed previously. Someone else in the room now approaches the front of my desk. I might propose one of two topics for conversation: the yellow pad or the spot still present in my visual field. Which shall we discuss?

The metaphysical foundations for the spot and the pad are the same, and the transient nature of the spot does not matter essentially. But when I encounter this other sentient entity, we shall certainly have an easier time discussing the pad than the spot. Although she might well understand how a dark appearance might indeed be some vestigial side-effect of the biomechanics of visual processing and equally well imagine experiencing a similar phenomenon, the pad presents something altogether more tractable. While neither of us can know the other's interiority with respect to the phenomenal aspects of this object, we can make some effective assumptions that I believe amount to a very prosaic cognate of the covariance principle in physics. For example, I know that the yellow quadrilateral form that appears just to the right of my keyboard as I sit behind my desk will still appear yellow, but further down and now to the left of said keyboard from where my visitor is standing. In other words, there is something persistent that transforms in predictable ways with regard to our respective experiences of this particular item from the world's vast inventory. A principle of symmetry operates here – not in the geometric sense, for the pad has a different aspect depending on where I stand – just as it does in science, to justify an objective ontological posit. That is why we can so easily agree to talk about the pad, and not so easily about the spot. The kind of symmetry that would justify discourse with respect to the latter is much more elusive: the pad has an obvious aspect of covariance, while the spot

processive flux against an invariant background of unchanging forms. Stolz claims that Whitehead had implicitly imported these ideas from science into his process metaphysics to develop a system that is "perspective invariant". Thus Stolz speaks of process thought as a "covariant natural philosophy".

does not. Indeed, to someone who knew nothing about the physiology of vision and had never experienced after-images herself, a conversation about the spot might prove quite amusing or frustrating, depending on one's temperament.

One way of putting this, then, is that an objective ontology countenances that which remains invariant about the world as we shift from subject to subject. We apply a certain class of abstract symmetries to our experience under which the objects of ordinary discourse remain invariant. These symmetries exist by virtue of the existence of a world independent of its experiencing entities and at the same time are the basis for our coherent discourse. Thus the correspondence and coherence theories of truth are aspects of both the world as given and our reconstruction of it as based upon direct experience. In many cases, the common transformations that reconcile reports from various subjects override direct perception. When I say that my pad is rectangular, even though it does not appear so, I am speaking of a property it holds by virtue of a putative shared coordinate system. Indeed, I think that this way of speaking is a key element of Whitehead's famous fallacy of misplaced concreteness.

While we argue here that an objective worldview is built out of a generalized notion of covariance, we emphasize that in another sense, any such view depends essentially on a kernel of subjectivity that defeats transformation: the private, unsharable, incorrigible interiority that is experience. To give a fairly representative example of this, consider what, in pre-relativity days, we would have thought of as a perspective-free invariant of a somewhat regular physical object: its length. (Here we again have the reciprocity between objecthood and covariance.) Using a ruler that can be read easily to a sixteenth of an inch, we measure the length, say, of an envelope. The edge of the envelope falls conveniently on one of the ruler's tick marks. I make the measurement and announce the result, and then my lab associate does likewise. Where is the objectivity in this process? The report is perspective-free, but not so in the least is the experience of reading the ruler.

5.1. Some classifications by symmetries

A physical law satisfies the covariance principle if its form is preserved by certain admissible shifts in coordinate systems. Hence in science, covariance is relative to a class of transformations. In our metaphysical extension, this generalized notion of covariance and its attendant consequences for ontology are likewise relative to classes of perspective shifts that correspond to certain general domains of discourse. For example, physical objects are largely defined by symmetries in location, duration, extent, composition, and certain derivatives thereof. Mathematical objects depend similarly on (the symmetries of) identity, cardinality, extent, certain formal notions of relatedness and derivatives thereof. Indeed, the objects of science and mathematics seem so knowable exactly because their defining symmetries are so sharp. What, for instance, could be more perspective-free than the cardinality of a finite set? How, then, could two finite sets of different cardinality be identical?

Can there be other kinds of world objects? Let us consider for the moment the psychological construct of *hunger*, or more simply an instance of hunger, as we might consider an instance of a physical object, like a book. To put our connection between covariance and ontology to the test, we should ask if there is some symmetry, some perspective shift, associated with such a usage. The answer, I would maintain, is yes, and this brings us to a whole new class of invariants.

I watch the actions, perhaps by video, of a person over the course of many hours. He works diligently in his home at some task, noting the time only infrequently. Then he seems distracted and looks again at his watch. He rubs his chin, looks down at his work, looks up again, inhaling as if he has reached some decision. He finally rises, walks to his kitchen, opens the refrigerator, then the pantry, peers here and there, moving, assessing, rejecting one item or another. I ask the basic question of covariance: what would *I* be experiencing if I were he, if *I* were acting this way? But, of course: he's hungry. I reach this conclusion easily enough by projecting myself into his narrative.

Let me stretch this hypothetical another length or two to by introducing a friend of mine, who has been watching the video with me. We freeze one frame in which we have a clear view of the interior of his refrigerator, and, in particular, of that common custom-molded shelf which is used to store eggs. It would be almost inconceivably unlikely that my friend and I, who have always gotten along easily enough with the world of ordinary events and discourse, could not agree on how many eggs were stored on that shelf. Yet when, near the end of the tape, I observe, "He's hungry", she shakes her head and says, "No, he's looking for something – something valuable that he's just realized he's misplaced". Again unlikely, but not enough to have me question the foundations of language or reality, or to suggest that my companion have her head examined. It might turn out, for instance, that she knows he's a diamond smuggler. Indeed, Hollywood has sometimes used more bizarre plot points.

The long point of this example is that matters of agency and intentionality may also find their way into objectified discourse via this more narrative version of covariance. The symmetries here depend on the underlying experiences in a different sort of way, which, I think, with sufficient analysis could be part of a well defined scheme. Without attempting any such deep analysis, let me, nonetheless, make a few plausible observations.

Note that counting the eggs in our hypothetical subject's refrigerator is an experientially flat operation. The relations involved are only identity and non-identity of physical objects, *qua* physical objects, and the notion of pairing distinct objects with number signs. My point is that the experiences that underlie the assessment of cardinality are largely horizontal: I need not account for the history of the eggs or the manufacture of the refrigerator in making this assessment. The assessment of hunger, on the other hand, has a much richer, and at least a two-dimensional experiential base. It is possible, but much less likely that we might conclude that someone is hungry from a still photograph. The complexity of such assessments, of course, can grow without apparent limit, but this kind of covariance is what makes narratives intelligible to us. It accounts for how stories that did not touch at the age of six, can sometimes begin to reach us at the age of sixteen, and only be fully understood much later in life. It accounts for how we finally come to understand that Mr. Darcy in fact loves Eliza in *Pride and Prejudice*. It accounts for why there are so many prodigies in mathematics and physics, and so few, if any, in history and literature.

Although mathematics, physics, history and literature can be said to emphasize different kinds of symmetries that ultimately account for the different epistemological and ontological feels of these disciplines, we still must stress, that there is nonetheless an underlying unity of process, and the foundations in all cases are the atoms of unshared and unsharable experience. What are shared and sharable by virtue of the nature of reality are *relations*. I don't know what it feels like to be the famous neuroscientist Mary (Jackson 1982) when she finally sees red, and she doesn't know what it feels like to be me when I see red, and so, too, for green, but the relations between red and green will be preserved well enough for us to use the words intelligibly. A distinction sometimes made in this connection is between the psychological and the phenomenological. One might say that Mary and I can only discuss red and green as a matter of psychology, not phenomenology. Ultimately, then, covariance, while founded on direct experience, is about relationships, and that is how so many of us can sometimes agree that we love the same symphonies and abstract paintings.

6. Covariance, process metaphysics, and evolution

So far we have considered how we determine the objects in the world based on unsharable experiences. For us as individuals, the objects arise out of groupings of experiences along the lines recognized in mathematics as a quotient construction. For us as a language-capable community, our shared objects of discourse answer to a metaphysical analog of the general covariance principle in physics. This is a matter of respect for the symmetries of the perspectives belonging to various subjects. One might want to argue that nothing has been said here that could not be said without the idea of covariance, but as I remarked in connection with the borrowing of the quotient construction into metaphysics, having a well thought out technical model as the basis for a structural metaphor does enhance the likelihood of its ultimate coherence.

I mentioned in footnote 12 that Joachim Stolz has defended an extended thesis that process theory should be considered a "covariant natural philosophy". While my approach to covariance has been more abstractly mathematical and correspondingly more general than that of Stolz, I do want to emphasize some basic consonances with both Whitehead and Stolz's interpretation of Whitehead.

First, the basis for ontological construction at both the individual and community levels consists of the real integrative experiences that are constitutive of the actual entities of the world. This is, of course, essential to Whitehead's notion of concrescence. By ontological construction, I mean in both cases a passing from the subjective to objective worldview with, again, no presumption of idealism. What is presumed is a threshold level of development in the subject that supports, at the very least, temporal and spatial integration of experience. What the lower limits of this threshold might be, we cannot yet say, although – as I shall suggest below – this is not necessarily a matter of scientific advance as we now understand science.

Second, and implicit in the preceding point, is that this worldview is fundamentally relational, not just at the level of concrescence, but all the way up. Private objects are given by the private relations among experiences that define the requisite equivalences. Public objects are given by the public – or perhaps more properly, intersubjective – relations of the requisite covariance. Further, the notions of private and public, as we have suggested above, are not entirely dichotomistic. There is a spectrum, toward the middle of which we have a kind of narrative covariance built out of relations among phenomena that do not pass directly into the public domain; matters of intentionality versus matters of cardinality, for example.

Third, we have Stolz's thesis in *Whitehead und Einstein* that at the ground level process thought is covariant in the most fundamental sense of all: the world process has the same formal description for *all* actual occasions, and this holds, loosely speaking, whatever their mutual temporal and spatial relations may be. I say, "loosely speaking", because the notions of time and space as understood by science do not properly apply in the domain of metaphysics. In this sense, what Stolz has so astutely observed about Whitehead is that he, in his time, forced a scientific principle that had only recently emerged – a scant decade before *Process and Reality* took shape – back into the proto-scientific discourse of metaphysics.

There is, of course, a huge intentional difference between asserting that some principle is generative of the world as opposed to merely descriptive. Perhaps the most famous case of this occurs with regard to science itself insofar as Thomas Kuhn made his famous charge that science is both descriptive and constitutive of reality. I want to speculate now on whether covariance can be said to be an active organizing principle in the development of the world, and for reasons mentioned above, I think that in this, unlike Kuhn, I am in no way flirting with idealism.

A first and limited case for this hypothesis can be made at the intermediate level of the world – that which lies between the microcosmic and macrocosmic scales – in connection with Darwinism. In the context of the ordinary world of physics and biology, some sentient entities will encounter other sentient entities. Whatever capacity these have singly to piece together the world to their ultimate benefit will naturally be enhanced and extended to the extent that they pass into a community-based description.[13] This depends on some given element of coherence provided by whatever cognitive aspects are shared across individuals (shades, again, of Nagel, and this time of his celebrated bat) and on the acceptance of some operating principle that allows for descriptions to make sense across perspectives; in other words, a species of covariance.[14] In this sense, covariance, like science and language, might be seen as an aspect of a larger, engulfing dynamical principle, which in this case is Darwinism.

A more radical and powerful take on all of this is the reciprocal possibility that Darwinism is somehow subsumed by covariance, and here another element of Whitehead is paramount. Consider his deep thesis that the evolution of reality seems to bring about experiences of greater and greater intensity. This, I hasten to add, is manifestly not a direct matter of survival, since those of us who believe that mechanical systems have no consciousness can still envision some pretty rugged ones. In fact, the central thesis of epiphenomenalism, taken seriously, is that experience doesn't matter to the

[13] The efficacy of public descriptions comes, of course, at a cost to our individual narrative constructions. But to comment further here would be redundant: the issue has been sublimely explicated by Jane Austen in *Sense and Sensibility*.

[14] This brings me to the point anticipated above on the limitations of science as we now understand the term. Nagel's bat reminds us that while we can presume that *it feels like something* to be an entirely different kind of creature, that feeling may be forever beyond our capacity to share in any significant way. (See also his less well known spider in Chapter XI of *The View from Nowhere*.) This implies, in the context of our current discussion, that we may never complete the science of bats: to whatever extent such science ultimately converges, it misses something that is there in the world even in the public sense that defines the scientific domain. I can only describe this missing element as "the science that bats might themselves do".

ultimate trajectory of the world.[15] Now the only plausible mechanism for intensification that I can imagine would implicate some sort of coherence, which is to say a distinguished species of amalgamation analogous to physical resonances. The general idea then would be a kind of experience augmented dynamically, sharpened and more discriminating by virtue of being integrated across a myriad of component entities. This, I think, has two implications. First, what is made coherent must in some prior sense be sharable. But experience is by its nature private and interior, so this sharability must assimilate to a properly distinct notion that reconciles the perspective inherent in all experience with the possibility of perspective invariance. This notion can be properly called covariance. Second, since, just as in physics, the idea of covariance presupposes the existence of common forms, any metaphysical system that deeply implicates covariance cannot be supervenient on its physical basis.

This second point goes beyond the obvious observation that any system that speaks of perspective and experience must be at least epiphenomenalistic insofar as it acknowledges experience as part of the world. But orthodox epiphenomenalism is further distinguished by the assumption of physical closure, and beyond this even, the assertion that the world, *including its experiences*, is supervenient on its physical base. Thus the existence and efficacy of common forms in affecting the direction of a world based on a covariant metaphysics would clearly transgress one of the defining boundaries of epiphenomenalism. Hence we are led to the kind of dual aspect theory that I have suggested – but only in the abstract – elsewhere (Valenza 2008b, 2010). The point is that the broadest principles that govern the evolution of the world as a unified process that encompasses both the uncompromised interiority of experience and physical objectification include certain privileged formalisms that, deliberately to echo Kuhn, are both descriptive and constitutive of the process. Those formalisms then express themselves in metaphysical covariance and accordingly become an aspect of the world's evolution in the phylogenic sense. Looked at in this way, the survival dynamics of covariance as an attribute of an effective basis for community-shared truths may be a consequence of this deeper property of the world and not an explanatory cause. Using one of the classic

[15] See Chalmers (Chapter 5, 1996) on "the paradox of phenomenal judgment".

texts in theoretical biology as a springboard, we expand upon this idea in the following section.

7. The evolution of physical and experiential forms

The ninth chapter of D'Arcy Thompson's book *On the Growth of Form* (1917) is entitled "On the Theory of Transformations, or on the Comparison of Related Forms". In this section we summarize the key idea of this chapter and put it into the context of this paper, with significant extension. The relevant contextual term from the lexicon of contemporary mathematics is *homotopy.*[16, 17]

Roughly speaking, two shapes (more technically, topological spaces) are said to be homotopic if each can be continuously deformed into the other. Thus the respective surfaces of a pea and a peanut are homotopic, as are those of an apple and a pear, but not those of an apple and a donut. The idea of homotopy is very broad, as illustrated by the two-dimensional shapes in the figure below.

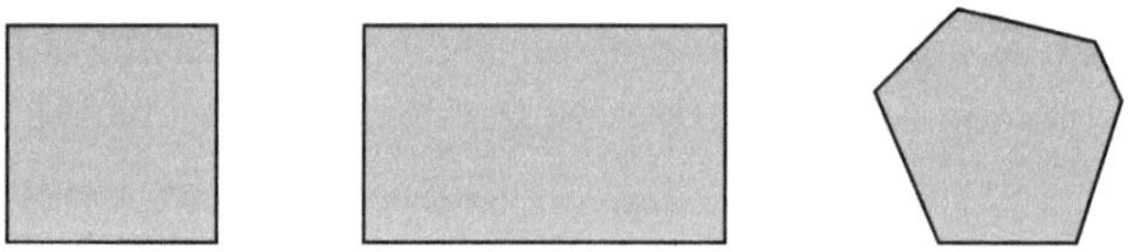

Fig. 7: Some homotopic shapes

It is easy to imagine that each of these three shapes can be continuously deformed into one another, but clearly not with equal felicity. Thus to deform the square on the left into the rectangle in the middle, we need only stretch horizontally, but the deformation of the square into the irregular hexagon is not nearly as straightforward. If we think of these shapes as embedded in a standard two-dimensional coordinate system, an obvious way to capture

[16] I gratefully acknowledge that the inspiration for connecting the work of D'Arcy Thompson with the themes of this paper was the result of a suggestion made to me by Spyridon Koutroufinis, editor of this volume.

[17] *Homology* is a related term used in both mathematics and biology, but due consideration of Thompson's point of view strongly suggests that here homotopy is more apt.

the simplicity of the former deformation is to note that we could specify the required transformation with a single parameter: the stretching factor to apply to the x-coordinate.

Thompson's point in his analysis of living forms is that body plans often exhibit the kind of homotopy illustrated by the square and the rectangle above; this is to say, of a sort characterized by a rather simply defined – or at least simply parameterized – transformation, as he illustrates with abundant examples. Note well that, as we have mentioned earlier in this paper, such transformations may equally well be considered as applied to the shape itself or to the coordinate system in which it is embedded. Thus the rectangle is equally well realized as a horizontally stretched square, or as the graph of an ordinary square in a horizontally stretched coordinate system. The view we take is mathematically neutral, but when we regard the objects as under transformation, we are disposed to condense our ontology by the quotient operation; in contrast, under coordinate transformations, we are disposed to filter our ontology by covariance or symmetry. In this light, it is delightful to observe that the typological definition of a species is actually an instance of the quotient operation.

What, then, shall we say about Thompson's observation? Since he uses the language of coordinate transforms, we may take as a consequence of his assertions that viable body plans exhibit covariance. One might further want to say that this reflects the covariance of nature, for instance in the cases of approximate bilateral or radial symmetry. Thus left-right symmetry is likely the norm for many species because on the surface of an approximately spherical planet, neither direction is preferred. In contrast, up-down asymmetry reflects the directionality of gravity. But this reflection of environmental symmetry in biological forms is limited: Why, for instance, do most creatures not manifest full horizontal radial symmetry, like a disk? Perhaps because it is more efficient to provide for a preferred direction of motion. But insofar as body plan symmetry or covariance reflects physical symmetry or covariance, we may illustrate the Darwinian picture with the following schematic:

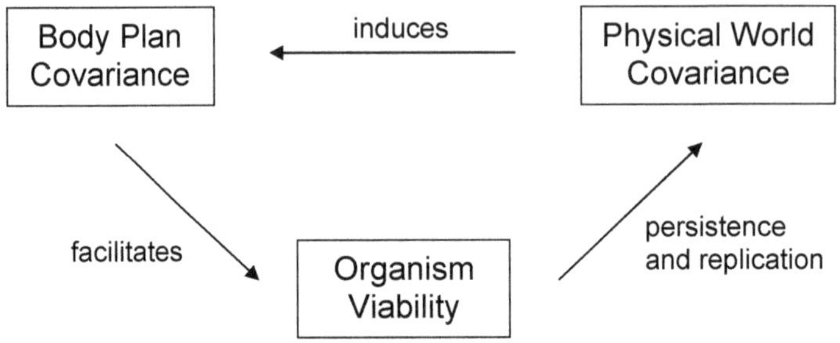

Fig. 8: A schematic for the evolution of physical forms

The previous point is perhaps another facet of the general observation from the previous section about covariance as an aspect of Darwinism. The deeper question goes to the experiential or even cognitive nature of living things. Does covariance play a similar role here? The interpretation of Thompson's thesis in the physical domain suggests that that we might attempt to formulate a tripartite relation of similar structure with regard to experience. Fortunately, the necessary components are both provided by process metaphysics, as shown below.

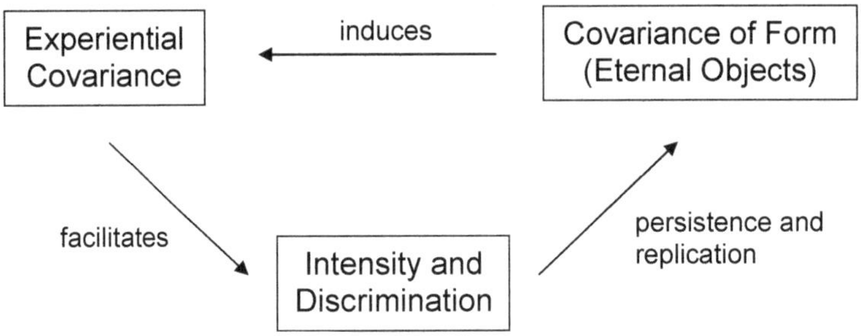

Fig. 9: A schematic for the evolution of experiential forms

The point of the figure above is two-fold. First, in attempting to mime the structure of the figure previous, we need an objective basis for experiential covariance, just as we have an objective basis for body-plan symmetry in the general symmetries of physics. This "covariance of form", as we here designate it, is for process metaphysics grounded in the eternal objects, which in their universality present the same aspect to all actual entities. Second, there is the ostensibly teleological aspect of body-plan symmetries insofar as they facilitate the survival of individuals and, consequently, of species. In the present structure, this aspect is provided by the Whiteheadian imperative toward entities capable of greater intensity and discrimination.[18] In both diagrams, the teleological aspect directly supports persistence and replication, in the first case of organisms, and in the second case of coherent instances of subjective forms in the concrescences of actual entities. Indeed, just as certain physical forms, high-level biomechanical structures, evolve, so do *experiential forms*, which emerge as the structures of consciousness.[19] Rationality thus becomes a feature of evolution in this broader sense. Accordingly, I think it is fair to say that these last two diagrams taken together as a speculative picture of the dynamics of evolution in both the physical and experiential domains – which is to say, evolution in a dual-aspect metaphysical theory, such as Whitehead's – both subsume and transcend the ordinary Darwinian paradigm.

Given the daunting complexity of the full Whiteheadian model, one might ask in the abstract just what is needed to support the dynamical relationships suggested by Fig. 9. I cannot answer this question here, except to make a few elementary observations using language introduced by David Ray Griffin (see his paper in this volume). We certainly do transgress the realm of *sensationist-atheistic-materialist naturalism* insofar as we grant at least some abstract formalisms causal efficacy, thus violating both sensationism and materialism. I am not at all sure, however, that this logically commits us to any species of theism, or, in particular, to the naturalism of process metaphysics, which Griffin characterizes as *prehensive*, *panentheistic* and *panexperientialist*. Nonetheless, I do suspect that admitting any kind of awareness of form into our metaphysics will take us a long way toward Whitehead.

---

[18] Whitehead would be more likely to speak of *contrast* rather than discrimination, but I prefer this more aesthetically charged term.

[19] In the title of this section, I would have preferred to speak of the evolution of physical and *subjective* forms rather than physical and *experiential* forms, but, as we see here, the term *subjective form* has been co-opted by Whitehead to designate a technical aspect of concrescence.

8. Conclusion

The role of covariance in its general philosophical sense in our common habits of thinking about the world and even measuring its progress is unquestionable, not just in science but in its political organization and elsewhere.[20] In the last three millennia, in moving from Plato's *Republic* to Hobbes's *Leviathan* through the Enlightenment toward democracies in which the vote has been extended to virtually the entire adult citizenry, truth is no longer only to be seen from one privileged perspective – that of the philosopher-king, for instance – but is now for everybody, or at least many of us would so hope. This is indeed a principle of covariance operating in political philosophy. The arguments above tend to pull us away from the position that this is some sort of generalized Darwinian dynamic. One would say rather that greater and greater complexity in physical, cognitive, and perhaps even social objects answers to something more fundamental. Indeed, the point made by John Cobb, again in this volume, that the activity of living beings plays a role in evolution (his "fourth variable") is a species of the general scheme we have been discussing here exactly to the extent that the activities of a subject include its experiences as much as its physical states.

With respect to Whitehead, we have seen that his ideas are again enormously consonant with this framework. Darwinism essentially ends with an effective means to survive the world's hazards into reproductive maturity, and of course such means include effective representations of the world. It cannot by definition account for organizing forces external to physical reality. But if covariance is an expression of privileged forms in the sense that I have been speaking, something more is required, and Whitehead's fundamental ideas of eternal objects and God meet this requirement nicely: God is the realization of the forms requisite to any covariant metaphysics, and God acts as lure in the creative process so that these shared forms are given expression, and contrast and intensity. The long history of increasing complexity of the jointly physical and psychical processes would, from a

[20] It has been observed that Schönberg's atonality was an explicit transfer of Einstein's general covariance into the domain of music. The point is that the privileged role played by certain notes and chords in the context of a key is thereby summarily dissolved.

Whiteheadian perspective, flow from this. Indeed, the singularly swift evolution of *living* things, things distinguished moment by moment by their potential for creative response to the antecedent, objectified world, becomes far more comprehensible in this view.

REFERENCES

Chalmers, D. (1996). *The Conscious Mind.* New York: Oxford University Press.

Eilenberg, S.; Mac Lane, S. (1945). "General Theory of Natural Equivalences". In: *Trans. Amer. Math. Soc.*, 59, pp. 231-294.

Henry, G.; Valenza, R. (2001). "Eternal Objects at Sea". In: *Process Studies*, 30/2, pp. 55-77.

Jackson, F. (1982). "Epiphenomenal Qualia". In: *The Philosophical Quarterly*, 32, pp. 127-136.

Nagel, T. (1986). *The View from Nowhere*. New York: Oxford University Press.

Lindsay, R.; Margenau, H. (1963). *The Foundation of Physics.* New York: Dover Publications.

Munitz, M. (ed.) (1971). *Identity and Individuation*. New York: New York University Press.

Resnick, R. (1968). *Introduction to Special Relativity*. New York: John Wiley & Sons.

Stolz, J. (1995). *Whitehead und Einstein*. Frankfurt am Main: Peter Lang.

Thompson, D'Arcy (1992). *On Growth and Form*. Cambridge: Cambridge University Press. (Abridged edition edited by John Tyler Bonner, Canto edition, 1961; original unabridged edition, 1917)

Valenza, R. (2008a). "Albert Einstein". In: Weber, M.; Desmond, W. (eds.). *Handbook of Whiteheadian Process Thought*, Volume 2, Frankfurt: Ontos Verlag, pp. 400-408.

—— (2008b). "Introduction to 'The Metaphysics of Consciousness and Evolution'". In: Cobb, J. (ed.). *Back to Darwin. A Richer Account of Evolution*. Grand Rapids, MI: Eerdmans Publishing Company, pp. 263-267.

—— (2010). "Metaphysical Models". In: *Process Studies*, 39 (1), pp. 59-86.

Wiggins, D. (1980). *Sameness and Substance*. Cambridge: Harvard University Press.

Index

Bei Fragen zur Produktsicherheit wenden Sie sich bitte an:
If you have any questions regarding product safety,
please contact:

Walter de Gruyter GmbH
Genthiner Straße 13
10785 Berlin
productsafety@degruyterbrill.com